Glucose-sensing Receptor in Pancreatic Beta-cells

Itaru Kojima
Editor

Glucose-sensing Receptor in Pancreatic Beta-cells

Springer

Editor
Itaru Kojima
Institute for Molecular & Cellular Regulation
Gunma University
Maebashi, Gunma, Japan

ISBN 978-981-13-0001-1 ISBN 978-981-13-0002-8 (eBook)
https://doi.org/10.1007/978-981-13-0002-8

Library of Congress Control Number: 2018942892

Printed on acid-free paper

This Springer imprint is published by the registered company Springer Nature Singapore Pte Ltd.
The registered company address is: 152 Beach Road, #21-01/04 Gateway East, Singapore 189721, Singapore

Contents

1 **History of the Research on the Glucose Receptor** 1
Willy J. Malaisse

2 **Cell-Surface Glucoreceptor Recognizing Anomers of Glucose in Pancreatic β-Cells** 29
Yuko Nakagawa and Hatsumi Niki

3 **KATP Channel-Independent Pathway and the Glucoreceptor** 37
Toru Aizawa and Mitsuhisa Komatsu

4 **Signaling System Activated by the Glucose-Sensing Receptor** 51
Johan Medina and Yuko Nakagawa

5 **The Role of the Glucose-Sensing Receptor in Glucose-Induced Insulin Secretion in Pancreatic β-Cells** 61
Yuko Nakagawa and Johan Medina

Chapter 1
History of the Research on the Glucose Receptor

Willy J. Malaisse

Abstract The history of the research on the glucose receptor represents the main matter of the present chapter. A short introduction mentions two reviews published in 1972 and introduces the concept of a pancreatic islet B-cell glucoreceptor. Attention is then drawn to the anomeric specificity of glucose-stimulated insulin secretion and its possible metabolic determinants. The verbal care required for the use of the glucoreceptor word is duly underlined. Examples of the modulation of key enzyme activity in insulin-producing pancreatic islet cells by D-glucose itself or one of its metabolites are then provided in the framework of the interaction of phosphorylase a with D-glucose anomers, the activation of glucokinase by a fructose-1-phosphate-sensitive regulatory protein, and the activation of phosphofructokinase by fructose-2,6-bisphosphate. The riddle of L-glucose pentaacetate insulinotropic action, the effects of artificial sweeteners on insulin release, and last but not least, the presence of the sweet taste TIR3 receptor in pancreatic insulin-producing islet cells are eventually and duly also considered.

Keywords Pancreatic islet · B-cell glucoreceptor · Anomeric specificity of D-glucose insulinotropic action · Glucoreceptor myth · Interaction of phosphorylase a with D-glucose · Activation of glucokinase by a regulatory protein · Activation of phosphofructokinase by fructose-2,6 bisphosphate · L-Glucose pentaacetate · Artificial sweeteners · TIR3 receptor

1.1 Introduction

The major aim of the present chapter is to evoke, in a historical perspective, the development of distinct concepts on the process of glucose recognition by pancreatic islet beta cells as a stimulus for insulin release.

W. J. Malaisse (✉)
Department of Biochemistry, Brussels Free University, Brussels, Belgium
e-mail: malaisse@ulb.ac.be

I. Kojima (ed.), *Glucose-sensing Receptor in Pancreatic Beta-cells*,
https://doi.org/10.1007/978-981-13-0002-8_1

Already in 1969, at the occasion of the 13th Nobel Symposium, Cerasi and Luft postulated that glucose does not act solely as a substrate metabolized in insulin-producing cells but also, perhaps by allosteric effect, on a specific receptor which, in turn, activates adenyl cyclase [1]. In 1972, an excellent review by Randle and Hales was devoted to the glucoreceptor mechanism in the B-cell. These authors emphasized that two models could account for the glucoreceptor system: the first one based on the combination of glucose itself with a receptor molecule and the second one based on the generation of a signal as a product of glucose metabolism in the B-cell [2].

In pages 199–214 of volume 1 ("Endocrine Pancreas") of the section 7 ("Endocrinology") of the *Handbook of Physiology* published in 1972 by the American Physiological Society, Frans M. Matschinsky contributed in the part of this textbook devoted to "Biochemical Organization of Islet Tissue" the 11th chapter of this volume under the title "Enzyme, metabolites, and cofactors involved in intermediary metabolism of islets of Langerhans." The author carefully underlined in the "conclusion" of his chapter that "whether metabolism of glucose is an absolute requirement for all stages of the multiphasic endocrine response evoked by glucose is disputed" and "that the two processes (glucose metabolism and insulin release) can exist independently of one another" [3]. He also drew attention to the fact that in contrast to the almost instantaneous secretory response to a rise in extracellular D-glucose concentration, the intracellular levels of most glycolytic intermediates only rise relatively slowly, indicating that only the late phase of insulin release might be controlled by the metabolism of the hexose. It was speculated that metabolites may modify the responsiveness of β-cell receptor(s) of the hexose. Such a receptor hypothesis was rapidly and largely accepted, in part maybe because of its simplicity, by scientists interested in the issue of glucose recognition by the β-cell as an insulin secretagogue.

1.2 The Anomeric Specificity of Glucose-Stimulated Insulin Secretion

At about the same time, in 1974, the most crucial finding, in my opinion, on the issue of glucose recognition by the insulin-producing β-cell, was reported by A. Niki, H. Niki, I Miwa, and J. Okuda in an article published in *Science*, entitled "Insulin secretion by anomers of D-glucose" and documenting that the α-anomer of D-glucose is better able than the β-anomer to stimulate insulin secretion [4]. This finding was then confirmed, with an amazing rapidity in experiments all conducted in 1974 by Grodsky et al. [5], Rossini et al. [6], Matschinsky et al. [7], and Idahl et al. [8]. It was also shown that α-D-glucose was more potent than β-D-glucose in suppressing glucagon secretion [6, 7, 9], inducing a rapid transient efflux of phosphate from perifused islets [10], provoking cyclic AMP accumulation in rat islets [11], and protecting β-cell against the cytotoxic action of alloxan [12, 13]. The conclusions of these studies were that beta cells distinguish the α and β anomers of D-glucose for triggering insulin secretion at the receptor site of the cell membrane

[4], that the anomeric specificity of the secretory response to D-glucose provides evidence for a glucoreceptor, the action of glucose being independent of intracellular glucose metabolism [5], that glucose acts on a cell membrane receptor to initiate insulin release [11], that the initial signal for stimulation of insulin release and inhibition of glucagon output is at the level of a glucoreceptor, independently of major pathways of glucose metabolism [9], and that both α-cells and β-cells contain glucoreceptor controlling glucagon and insulin secretion [7]. Rossini and colleagues, however, avoided speculations on a comparable issue. Last, Idahl et al. observed that β-D-glucose was at least as effective as α-D-glucose in stimulating counter-transport of 3-O-methyl-D-glucose, increasing the islet content of glucose-6-phosphate, and diluting $^{3}H_2O$ that results from the metabolism of D-[5-^{3}H]glucose, indicating that the anomeric specificity of the insulin-releasing D-glucose recognition systems is not shared by the earliest steps of glucose metabolism [8]. In fair agreement with the first of the three latter findings, Miwa et al. observed, using L-[1-^{14}C]glucose as an extracellular space marker, that over 5 min incubation at 37 °C, the uptake of β-D-[1-^{3}H]glucose by rat pancreatic islets was about twice higher than that of α-D-[1-^{3}H]glucose (16.7 mM each). It was concluded that the anomers of D-glucose each has a preferential function in pancreatic beta cells, α-D-glucose stimulating insulin secretion and β-D-glucose being transporter into the cells [14].

1.3 Alternative Hypothesis

Shortly thereafter, in 1976, in an article entitled "Identification of the α-stereospecific glucosensor in the pancreatic B-cell," it was first mentioned that the α-stereospecific system responsible for the greater ability of α-D-glucose, as distinct from β-D-glucose, to stimulate insulin secretion, to increase the concentration of cyclic AMP, to provoke the efflux of phosphate ions, and to suppress the release of glucagon in pancreatic islets remained unknown. It could be a membrane-associated glucoreceptor, a carrier for glucose transport across the cell membrane or an enzyme involved in the early steps of glucose metabolism [15]. The latter hypothesis was examined [15, 16].

No significant difference could be detected in the rate of α- and β-D-glucose phosphorylation by the islet enzymes. Indeed, over 5 min incubation in the presence of ATP (0.1 mM) and the islet homogenates, both anomers (9.0 mM) inhibited to the same degree the formation of [1-^{14}C]glucose-6-phosphate from [1-^{14}C]glucose (1.0 mM) at anomeric equilibrium. Moreover, when the islet homogenate was exposed for 6 min to [γ-^{32}P]ATP, the same amount of glucose-[6-^{32}P]phosphate accumulated whether in the presence of α- or β-D-glucose (10.0 mM).

The first indication of a difference in the metabolism of the two anomers was obtained by examining the affinity of the islet glucose-6-phosphate dehydrogenase towards α- and β-glucose-6-phosphate. The hexose-phosphate was extemporaneously generated from each anomer in the presence of yeast hexokinase and ATP. When the generation of glucose-6-phosphate was the rate-limiting factor, the formation of 6-phosphogluconate by the islet homogenate occurred later and at a slower rate in the presence of α- as distinct from β-D-glucose. Since yeast hexokinase does not act more effectively on β- than α-D-glucose, the experimental results indicated that in islets, like in other tissues, the enzyme glucose-6-phosphate dehydrogenase is stereospecific for β-D-glucose-6-phosphate. Also more sorbitol accumulated in islets exposed for 5 min to freshly dissolved β- and distinct from α-D-glucose (16.7 mM). This behavior, which could theoretically be due to a stereospecific affinity of the islet aldose reductase, might well result from the preferential orientation of β-D-glucose-6-phosphate to the pentose pathway, more NADPH being then available for the conversion of glucose to sorbitol.

Since neither the phosphorylation of glucose nor its conversion to either 6-phosphogluconate or sorbitol offered a satisfactory explanation for the more marked insulinotropic action of α-D-glucose, attention was drawn to the possible participation of glycolysis in such a process. After 5 min incubation in the presence of D-glucose (7.2 mM), the concentration of glucose-6-phosphate was lower and that of further glycolytic intermediates higher in the islets exposed to α- as distinct from β-D-glucose. This suggested that the phosphoglucose isomerase of the islets, like that of other tissues, is stereospecific for α-D-glucose-6-phosphate. Such a view was confirmed, using a model comparable to that defined above for the study of the islet glucose-6-phosphate dehydrogenase. When the availability of glucose-6-phosphate was the rate-limiting factor (namely, at low glucose and yeast hexokinase levels), the rate of fructose-6-phosphate formation by the islet homogenate was almost 50 percent higher with α- than with β-D-glucose-6-phosphate.

Two independent findings indicated that in the intact B-cell like in the islet homogenate, the rate of glycolysis was higher in the case of α-D-glucose. First, in islets exposed for 6 min to [U-^{14}C]glucose (2.2 mM) in anomeric equilibrium, much less radioactivity was recovered in $^{14}CO_2$ when unlabeled α- as distinct from β-D-glucose (7.8 mM) was also present in the incubation medium. This indicated that the α-anomer is better able to dilute the metabolic pool from which $^{14}CO_2$ is eventually derived, under conditions where more than 90 percent of the total $^{14}CO_2$ production is accounted by glycolysis. Second, the output of lactate from islets incubated for 5 min with freshly dissolved α-D-glucose (7.4 mM) was significantly higher than that found in islets exposed to β-D-glucose.

In islets prelabelled with ^{45}Ca and perifused in the absence of extracellular Ca^{2+} and presence of EGTA (1.0 mM), the inhibitory effect of D-glucose (5.8 mM) upon ^{45}Ca efflux was more pronounced in the case of α-D-glucose as distinct from β-D-glucose. Consistent with the latter finding, the glucose-induced increment in ^{45}Ca net uptake, above the basal value found in the absence of glucose, was significantly higher, after 6 min incubation in the presence of α- as distinct from β-D-glucose (8.4 mM). Over 6 min incubation, the α-anomer (8.1 mM) also provoked a higher release of insulin than the β-anomer.

Taken as a whole, these data indicated that the more marked insulinotropic action of α- as distinct from β-D-glucose is associated with a higher glycolytic flux, itself attributable to the stereospecificity of the islet phosphoglucoisomerase, eventually resulting in a higher accumulation of Ca^{2+} in the B-cell and, hence, a higher Ca^{2+} triggered insulin release. These findings were, therefore, considered to provide an essential support to the so-called fuel hypothesis for glucose-stimulated insulin secretion [16].

1.4 The Glucoreceptor Myth

An article entitled "Insulin release: the glucoreceptor myth" was even published in 1987 [17]. Its aim, however, was to underline the confusion arising from the current use, at that time, of an ill-defined glucoreceptor concept. Three acceptations of such a glucoreceptor were considered.

The first acceptation was inspired by the essential debate concerning the mode of identification of D-glucose as a stimulus for insulin release, which long opposed the supporters of a metabolic hypothesis and a glucoreceptor theory [16]. The former hypothesis postulates that the release of inulin is causally linked to an increase of D-glucose metabolism in the B-cell [16]. The latter theory postulates that D-glucose itself stimulates the B-cell through binding to a stereospecific glucoreceptor located in the B-cell, possibly at the level of the plasma membrane [4, 5].

When it was claimed that the secretory response of the B-cell to D-glucose is causally linked to an increase in D-glucose metabolism, the concept of a glucoreceptor, in its first acceptation, apparently became obsolete. The word glucoreceptor did not disappear, however, from the relevant literature. Instead, the same word became used, by those investigators who had defended the glucoreceptor concept in its first acceptation, to refer to an essential component, such as glucokinase, of the biochemical device responsible for the regulation of D-glucose metabolism in the B-cell [18, 19].

A third acceptation of the glucoreceptor was not ignored. Thus, without prejudice as to the intimate mode of action of D-glucose in the pancreatic islet B-cell, the latter could be considered to be equipped with a suitable glucoreceptor system. This would indeed represent a third acceptation of the word glucoreceptor. In certain cases of noninsulin-dependent diabetes mellitus, the insulinotropic action of D-glucose seems to be much more severely affected than that of other secretagogues [20]. Likewise, fasting causes a preferential decrease in the B-cell responsiveness to D-glucose [21]. In these situations, one could state that the function of the glucoreceptor system is apparently perturbed. This mode of expression was not uncommon. For instance, Garvey et al. observed that after a 3-week period of continuous subcutaneous insulin infusion to noninsulin-dependent diabetic subjects, the pancreatic insulin secretory function was significantly improved [22]. Hence, it was proposed that chronic hyperglycemia leads to impaired insulin secretion, "possibly by altering the function of beta cell glucoreceptors." For a reader of this interesting report, it was uneasy to decide which type of glucoreceptor the authors were actually referring to: a true receptor activated by D-glucose itself (first acceptation), a key enzyme of D-glucose catabolism (second acceptation) or the glucose-sensing function of the B-cell, whatever its precise nature (third acceptation). The decision may be important, since it is likely to condition the design of further experiments aiming at the identification of the molecular mechanism(s) responsible for this postulated phenomenon of "glucotoxicity."

Whenever the use of a given word represents a potential source of misunderstanding between scientists, it becomes imperative to clarify the issue in order to improve the communication of ideas. Is it possible to ascribe to the word glucoreceptor an unambiguous, and yet useful, meaning? In its first acceptation, it could seem adequate, if indeed solely used in such an acceptation. Otherwise, it was judged adequate, for all practical purposes, to banish the word glucoreceptor from reports dealing with the mechanism of glucose-induced insulin release. Each scientist would then have the freedom and responsibility to select an adequate word to convey his message. Terms such as glucose carrier or glucokinase could be used without ambiguity. If one wishes to refer to the system through which D-glucose is identified in the B-cell as a stimulus for insulin release, without taking position on the precise modality of such a recognition process, the expression "glucose-sensor device" could be recommended. Last, if it is thought that the mechanism of glucose identification coincides with the catabolism of the hexose in islet cells, why not use "glucose metabolism" when indeed referring to glucose metabolism?

The above proposal was considered, at that time, to offer a further advantage. Thus, if a true receptor, i.e., a molecule able to bind glucose and to generate, as a result of the binding process, a signal for insulin release other than a glucose metabolite, the word glucoreceptor could then be adequately reintroduced in our conception of the mechanism for glucose-stimulated insulin secretion. It was, at that time, considered unwise to dismiss such an eventuality [17].

As a matter of fact, in the same article entitled "Insulin release: the glucoreceptor myth," it was duly underlined that even in the framework of the fuel hypothesis for nutrient-stimulated insulin release, the concept of a receptor molecule for a given nutrient sexretagogue may nevertheless be adequate. For instance, the nonmetabolized analog of L-leucine, 2-aminobicyclo[2,2,1]heptane-2-carboxylic acid (BCH), stimulates insulin release apparently by causing allosteric activation of glutamate dehydrogenase [23, 24]. The latter enzyme could thus be viewed as a receptor for BCH [25].

Likewise, attention was drawn to the fact that the possibility remains that the D-glucose molecule itself, at the exclusion of its role as a substrate, modulates the activity of enzymes involved in the metabolism of carbohydrates in the islet cells. For instance, and by analogy with the situation observed in hepatocytes, it was considered as conceivable that D-glucose increases the rate of dephosphorylation of the active form of glycogen phosphorylase as catalyzed by phosphorylase phosphatase [17]. Three years later, such a hypothetical mechanism was indeed documented [26, 27], representing a fundamental aspect of the phenomenon referred to as an anomeric malaise.

The following sections of the present chapter provide further information in such a perspective.

1.5 Interaction of Phosphorylase *A* with D-Glucose Anomers

In 1977, insulin release due to glycogenolysis in glucose-deprived islets was first documented. When pancreatic islets were preincubated for 20 h in the presence of D-glucose (83.3 mM) and thereafter transferred to a glucose-free medium, theophylline (1.4 mM) provoked a dramatic stimulation of insulin release. This phenomenon did not occur when the islets were preincubated for either 20 h at low glucose concentration (5.6 mM) or only 30 min at the high glucose concentration (83.3 mM). The insulinotropic action of theophylline could not be attributed to contamination of the islets with exogenous glucose and was not suppressed by mannoheptulose. The secretory response to theophylline was an immediate phenomenon, but disappeared after 60 min of exposure to the drug. The release of insulin evoked by theophylline was abolished in calcium-depleted media containing EGTA. Theophylline enhanced the net uptake of ^{45}Ca by the islets. Glycogen accumulated in the islets during preincubation, as judged by both ultrastructural and biochemical criteria. Theophylline significantly increased the rate of glycogenolysis during the final incubation in the glucose-free medium. The theophylline-induced increase in glycogenolysis coincided with a higher rate of both lactate output and oxidation of endogenous ^{14}C-labelled substrates. These data suggested that stimulation of glycolysis from endogenous stores of glycogen is sufficient to provoke insulin release even in glucose-deprived islets, as if the binding of extracellular glucose to hypothetical plasma-membrane glucoreceptors is not an essential feature of the stimulus-secretion coupling process [26].

Much later, in 1990, attention was paid in the framework of the anomeric malaise to the anomeric specificity of the interaction of phosphorylase *a* with D-glucose. At very high concentrations (50 mM), D-glucose decreased the activity of muscle phosphorylase *a* by 85%, and the potency of the two anomers of the hexose was slightly higher with α- than β-D-glucose. The half maximally effective concentrations were clearly different: 4 mM with the α-anomer and 14 mM with the β-anomer, with intermediate value for the mixture of anomers at equilibrium, and about 25 mM for 1-deoxy-glucose, which at a very high concentration (50 mM) inhibited the enzyme by about 70%. The most striking difference between the two D-glucose anomers was evident at the lowest concentrations (1.0–2.5 mM), where β-D-glucose was barely inhibitor at all. The inhibition of liver phosphorylase *a* by the two glucose anomers was also investigated under the same conditions. The results were clearly similar to those recorded with muscle phosphorylase *a*. The anomeric specificity of the activation by D-glucose of phosphorylase phosphatase, as resulting from the binding of the hexose to the active phosphorylated *a*-form of phosphorylase, was also investigated. The rate of dephosphorylation of muscle phosphorylase *a* by protein phosphatase of the type-1 catalytic subunit was about 3.5-fold increased by 50 mM of either D-glucose anomers, and 50 mM 1-deoxyglucose was about as effective. Again, the half-maximally effective concentration was lower for the α-anomer of D-glucose (3.5 mM) than for the β-anomer (7.6 mM) and highest for 1-deoxyglucose (about 25 mM). The superiority of α-D-glucose was most pronounced at low sugar concentrations. This coincided with hyperbolic saturation kinetics with α-D-glucose, as distinct from markedly cooperative kinetics with β-D-glucose and 1-deoxyglucose. No obvious anomeric discrimination was observed, however, in the time-related (zero to 120 seconds) decrease of phosphorylase activity in glycogen-depleted rat hepatocytes exposed at 37 °C to D-glucose (5.0, 10.0 or 20.0 mM). The different response recorded in intact hepatocytes, as distinct from purified enzyme preparations, was eventually attributed to the rapid intracellular anomerization of D-glucose, itself accounted for by the interference of mutarotase, the half-life of D-glucose anomerization decreasing from 200 sec in cell-free media to 46 and 24 sec in intact cell suspension and sonicated hepatocytes, respectively [27].

Nevertheless, in glycogen-enriched rat pancreatic islets preincubated, when so required with D-[5-^{3}H]glucose, the mean values for both the inhibition of glycogenolysis by exogenous D-glucose (7.0 mM) and utilization of the exogenous hexose over 6 min incubation were higher in the presence of exogenous α- as compared to β-D-glucose [28].

Taken as a whole, these findings document not solely the effects of D-glucose itself upon the activity of selected enzymes catalyzing reactions involved in the intracellular fate of glycogen, but also the anomeric specificity of such effects. As such, the involved enzymes could thus be considered as a target for an anomeric specific response to the molecule of D-glucose.

1.6 Glucokinase-Catalyzed Phosphorylation of D-Glucose in Pancreatic Islets

In 1996, in a review article devoted to the metabolic signaling of glucose-induced insulin release, the four first sections concerned the role of glucokinase in such a process [29]. In the present conceptual context of the modulation of target enzymes activity by glucose or its metabolites, attention was drawn to the fact that islet cells are equipped with a glucokinase regulatory protein, which confers to glucokinase the property of being inhibited by D-fructose 6-phosphate and relieved from such an inhibition by D-fructose-1-phosphate. Taking advantage of the fact that pancreatic islets contain fructokinase [30], it had indeed been observed that in the postmicrosomal supernatant of pancreatic islets, prepared from fasted or fed rats, D-fructose 1-phosphate (10.0 mM) increases the activity of glucokinase by 20–30% as measured in the presence of D-glucose 6-phosphate and D-fructose 6-phosphate (4.0 mM). Further experiments documented the presence in pancreatic islets of a fructose-1-phosphate-sensitive inhibitor of liver glucokinase and the inhibition of islet glucokinase by the hepatic regulatory protein as assessed in the presence of 0.8 mM fructose-6-phosphates and the relief of such an inhibition by D-fructose 1-phosphate (1.0 mM). Advantage was further taken from the fact that fructose-1-phosphate can also be formed by condensation of D-glyceraldehyde with glycerone phosphate to document the presence of radioactive fructose 1-phosphate in islets incubated for 60 min in the presence of D-[U-^{14}C]glucose (10.7 mM) and D-glyceraldehyde (2.0 mM). It was proposed that the latter generation of D-fructose 1-phosphate may account for an increase in the rate of D-glucose phosphorylation and D-[5-^{3}H]glucose utilization caused by the triose in intact islets exposed to 10.0 mM D-glucose [31].

Thus, these findings indeed provided a second example of the modulation of the activity of a target enzyme, in this case glucokinase, by hexose metabolites in pancreatic islets.

Incidentally, in a study conducted in the postmicrosomal supernatant of rat liver homogenates, the phosphorylation of D-[U-^{14}C]glucose as measured in the presence of 5.0 mM ATP, 2.25 mM D-glucose 6-phosphate, and 0.75 mM D-fructose 6-phosphate indicated, by comparison of results recorded in the liver from either fed or 4-days starved rats, that starvation affected not only the temperature dependency and affinity for D-glucose of hepatic glucokinase, but also its responsiveness to D-fructose-1-phosphate [32].

The activation of glucokinase by its fructose-1-phosphate sensitive regulatory protein did not differ significantly, however, in the liver of control rats, Goto-Kakizaki rats with inherited noninsulin-dependent diabetes mellitus, and rats injected with streptozotocin during the neonatal period and, hence, considered as animals with acquired noninsulin-dependent diabetes [33]. To my knowledge, a comparable study was not yet conducted in pancreatic islets.

Two review articles were eventually devoted to the metabolic signaling of insulin secretion. The first of these articles was mainly concerned with the phosphorylation of glucose by both hexokinase and glucokinase. Emphasis was placed on the inhibition of hexokinase by endogenous D-glucose 6-phosphate and, to a lesser extent, D-glucose 1,6-bisphosphate in intact islets exposed to a high concentration of extracellular D-glucose, on the participation of cytosolic ATP in a synarchistic and sequential-type regulation of D-glucose phosphorylation in islets exposed to increasing concentration of the hexose, on the binding of hexokinase isoenzymes to mitochondria and its ambiguity susceptible to affect several metabolic variables in islet cells, on the intervention of the cytoplasmic regulatory protein conferring to glucokinase the property of being antagonistically regulated by D-fructose 6-phosphate and D-fructose 1-phosphate, and, last, on the possible changes in the intrinsic catalytic properties of glucokinase as conceivably resulting from the nonenzymatic glycation of cytosolic proteins [34]. The second review article drew attention to some current concepts concerning the organization of the β-cell glucose-sensing device. As such, the glucoreceptor myth, the glucokinase dogma, the regulation of D-glucose phosphorylation, the protein-to-protein interaction in metabolic processes, the regulation of D-glucose catabolism at sites distal to its phosphorylation, the coupling of metabolic to more distal events, the β-cell heterogeneity delusion, the physiopathology of metabolic signaling, the β-cell glucotoxicity, and desensitization misconceptions and therapeutic considerations represented the major issues under consideration [29].

1.7 Activation of Phosphofructokinase by Fructose 2,6-Bisphosphate

In the same conceptual context proposing that the activation of key glycolytic enzymes by selected hexose metabolites may correspond to a signal to receptor coupling process, it was documented that fructose 2,6-bisphosphate increases in a concentration-related manner (0.1–1.8 μM) the activity of phosphofructokinase in rat pancreatic islet homogenates incubated in the presence of fructose 6-phosphate (0.25 mM). The increment in velocity attributable to fructose 2,6-bisphosphate reached 50% of its maximal value at about 0.2 μM of the activator, the latter maximal value (recorded at 1.2 μM fructose 2,6-bisphosphate) averaging 337 ± 55 pmol. islet $^{-1}$.h^{-1}. At increasing concentrations of fructose 6-phosphate (0.1–5.0 mM) and in the absence of fructose 2,6-bisphosphate, the velocity of the reaction catalyzed by phosphofructokinase displayed, in semilogarithmic coordinates, a sigmoidal pattern, with half-maximal and maximal values close to 0.3 and 5.0 mM fructose 6-phosphate. In the presence of fructose 2,6-bisphosphate (1.2 μM), however, the relationship between reaction velocity and fructose 6-phosphate concentration in the same coordinates appeared hyperbolic with an apparent K_m for fructose 6-phosphate below 0.1 mM. When the islets were first incubated for 60 min in the absence or presence of glucose (20 mM) prior to homogenization, the ratio in phosphofructokinase activity at 0.25/5.0 mM fructose 6-phosphate averaged 40.5 ± 4.2 and 61.5 ± 6.8% ($n = 8$ in both cases, $p < 0.02$) in islets previously deprived of glucose and in islets first exposed to 20 mM D-glucose, respectively. These findings suggested that the glucose-induced stimulation of glycolysis in intact rat islets is attributable, in part at least, to activation of phosphofructokinase and that such an activation is mediated by fructose 2,6-bisphosphate [35].

By analogy with findings reported in July–August 1981 and revealing the presence in rat liver of a novel enzyme catalyzing the formation of fructose 2,6-bisphosphate from fructose 6-phosphate and ATP-Mg, an article, published in November 1981, documented the presence of fructose-6-phosphate,2-kinase in rat islet homogenates. The procedure used for such a purpose consisted of a first incubation of the islet homogenate in the presence of fructose-6-phosphate (5 mM) and ATP (5 mM) and, after alkali-treatment of the reaction mixture, of a second incubation in which the alkali-treated mixture was added to an assay cuvette containing purified phosphofructokinase and found to activate the latter enzyme in a manner comparable to that observed with fructose 2,6-bisphosphate [36].

In January 1982, a further article indicated that prior incubation of rat islets for 90 min in the presence (as distinct from absence) of D-glucose (16.7 mM) increased their content in an acid-labile activator of purified muscle phosphofructokinase, as judged from the effects of the alkali-treated homogenates upon the velocity of the reaction catalyzed by the purified muscle phosphofructokinase in the presence of fructose-6-phosphate (0.25 mM) and ATP (10.0 mM). The activity of islet fructose-6-phosphate,2-kinase could then be characterized in terms of its K_m for fructose-6-phosphate (0.08 mM), the absence of any detectable effect of glucose, and the close analogy between fructose-6-phosphate,2-kinase activity in liver (0.07 ± 0.02 mU/mg protein) and islets (0.06 mU/mg protein) [37].

One month later, in February 1982, a publication revealed (i) that prior incubation of rat pancreatic islets at increasing D-glucose concentrations (zero, 5.6 and 16.7 mM) causes a graded increase in the glucose-1,6-bisphosphate content of the islets, (ii) that glucose-1,6-bisphosphate both activates purified muscle phosphofructokinase and increases the velocity of the reaction catalyzed by phosphofructokinase in islet homogenates, (iii) that the combined effect of glucose-1,6-bisphosphate (4.0 and 8.0 μM) and fructose-2,6-bisphosphate (0.033 and 0.065 μM) upon muscle phosphofructokinase activity exceeded that of each activator tested separately, (iv) that the concentration of glucose-1,6-bisphosphate had to be approximately 100 times higher than that of fructose-2,6-bisphosphate in order to achieve a comparable activation of purified muscle phosphofructokinase, and (v) that, likewise, a 50 μM concentration of glucose-1,6-bisphosphate exerts effects upon islet phosphofructokinase activity comparable to those of only 1.0 μM fructose-2,6-bisphosphate. Nevertheless, the glucose-1,6-bisphosphate content of glucose-stimulated islets was considered sufficiently high to postulate that this hexose-bisphosphate participates, together with fructose-2,6-bisphosphate, in the activation of phosphofructokinase, with a resulting increase in both glycolytic flux and insulin secretory rate [38].

These findings were then the matter of a review article entitled "The glycolytic cascade in pancreatic islets" and published in 1982 [39].

A last publication revealed (i) that the increase of fructose-2,6-bisphosphate content provoked by exposure of islets to 20.0 mM extracellular D-glucose is much more rapid in purified pancreatic islet B-cells than in isolated hepatocytes and (ii) that in contrast again with the situation prevailing in the liver, glucagon fails to decrease the concentration of fructose 2,6-bisphosphate in either islets or purified B-cells. It was proposed that in the process of glucose-stimulated insulin secretion, an early increase in fructose 2,6-bisphosphate formation may, by causing activation of 6-phosphofructo-1-kinase, allow glycolysis to keep pace with the rate of glucose phosphorylation [40].

1.8 D-Glucose Metabolism in Pancreatic Islets

The fuel hypothesis for insulin release postulates that the process by which glucose is recognized as an insulinotropic agent entirely depends on the metabolic changes evoked by the sugar in the islet cells. The possible link between glycolysis and insulin release in isolated cells was reviewed in 1976 at the occasion of a meeting of the Minkowski Prize-Winners held in Capri [41]. The fuel hypothesis for insulin release was then proposed in 1979, in a review article published in Metabolism [42]. In 1981, the fact that D-glucose acts in the B-cell both as a substrate and enzyme activator was then proposed to permit reconciliation of the receptor and metabolic hypothesis [25]. Emphasis was eventually placed in 1983 on the possible coupling between the generation of second messengers (H^+, NAD(P)H, ATP) and more distal events in the secretory sequence, such as the remodeling of ionic fluxes across the plasma membrane and within the pancreatic B-cell [43].

Not surprisingly, attention was paid to several aspects of glucose metabolism in pancreatic islets, such as qualitative and quantitative aspects of glycolysis in isolated islets in 1976 [44], the influence of the environmental glucose concentration upon the production rate of metabolic end-products, and the intracellular concentration of both metabolites and co-factors in 1978 [45], the view that glucokinase could not be considered as the pancreatic B-cell glucoreceptor in 1985 [46], and the physiology and pathology of the pancreatic B-cell glucose-sensor device in 1990 [47], to cite only a few examples.

Since the anomeric specificity of D-glucose metabolism was considered as a key finding in support of such a fuel concept, attention was also paid to the metabolism of D-glucose anomers in rat pancreatic islets exposed to equilibrated D-glucose. Even in the islets exposed to D-glucose (2.8 and 8.3 mM) at anomeric equilibrium for 60 min at 4 °C, the metabolic fate of α-D-glucose differed vastly from that of β-D-glucose, the enzyme-to-enzyme channeling between hexokinase isoenzymes, especially glucokinase, and phosphoglucoisomerase being restricted to α-D-glucose 6-phosphate [48]. Under comparable experimental conditions, the β/α ratio for the conversion of D-[2-^{3}H]glucose to ^{3}HOH was higher than the β/α ratio for D-[5-^{3}H]glucose conversion to ^{3}HOH. These findings were considered as consistent with the postulated enzyme-to-enzyme tunneling of glycolytic intermediates between hexokinase isoenzyme(s), phosphoglucoisomerase, and possibly, phosphofructokinase [49].

1.9 The Riddle of L-Glucose Pentaacetate Action

Monosaccharides were and still are widely used as experimental tools in biomedical research, for example, as natural nutrients or specific inhibitors of carbohydrate metabolism. As a rule, their entry into cells is mediated by specific carrier systems. In a minireview reported in 1998 and entitled "Monosaccharide esters: new tools in biomedical research," attention was drawn to monosaccharide esters that are apparently able to cross the plasma membrane without requiring the intervention of such a transport system and then undergo intracellular hydrolysis in esterase-catalyzed reactions, so that the sugar moiety becomes readily available for further metabolism or metabolic action [50].

The conceptual background for the use of these monosaccharide esters consisted in the fact that, a few years previously, the esters of several carboxylic metabolites, such as succinic, glutamic, or pyruvic acids, had been proposed as tools for preventing ATP depletion in cells endangered by an imbalance between the synthesis and breakdown of this adenosine nucleotide [51]. The latter approach was inspired by the observation that such esters, e.g., succinic acid monomethyl or dimethyl esters, penetrate efficiently into various cell types in which they undergo intracellular hydrolysis. They are indeed better metabolized and exert more marked biological effects than the corresponding unesterified carboxylic molecules. For instance, succinic acid dimethyl ester is metabolized in rat pancreatic islets and, hence, stimulates insulin release, whereas succinic acid is virtually unable to enter islet cells and fails to display any insulinotropic action [52, 53]. In the light of these findings, it was considered that esters of monosaccharides may also be able to cross the plasma membrane and undergo intracellular hydrolysis, possibly resulting in a larger supply of their carbohydrate moieties than otherwise found in cells exposed to the corresponding unesterified sugars [50]. The key further sections of the latter minireview were dealing with pilot observations made with the α-anomer of D-glucose pentaacetate, the metabolic and functional effects of 6-O-acyl-D-glucose in rat erythrocytes and pancreatic islets, the fate of D-glucose pentaacetate in rat erythrocytes and its metabolism and insulinotropic action in rat pancreatic islets, and comparable information for L-glucose pentaacetate, D-mannose and D-fructose pentaacetates, D-galactose pentaacetate, D-glucose pentaacyl succinate esters, 2-deoxy-D-glucose tetraacetate, D-mannoheptulose hexaacetate, and streptozotocin tetraacetate [50].

The metabolic fate, cationic effects, and insulinotropic action of the polyacetate esters of nutrient and nonnutrient monosaccharides in rat pancreatic islets from normal and diabetic rats were then extensively investigated, being between 1997 and 2000 the matter of three scores of original articles and two further reviews [54, 55].

For the sake of briefness and in the spirit of the present chapter, further considerations are mainly restricted to L-glucose pentaacetate. An extensive study of β-L-glucose pentaacetate insulinotropic action provided the following information [56].

(i) *Phosphorylation of L-glucose*: L-glucose cannot act as a substrate for phosphorylation in the reactions catalyzed by pancreatic islet hexokinase isoenzymes.

(ii) *Hydrolysis of β-L-glucose pentaacetate in islet homogenates:* the hydrolysis of β-L-glucose pentaacetate in islet homogenates is about 4 times lower than that of α-D-glucose pentaacetate (0.25 mM each) [57].

(iii) *Uptake of β-L-[1-^{14}C]glucose pentaacetate by intact islets:* whilst the apparent distribution space of L-[1-^{14}C]glucose (2.0 mM) remained lower than that of ^{3}HOH, the apparent distribution space of β-L-[1-^{14}C]glucose pentaacetate (1.7 mM) largely exceeded that of ^{3}HOH, with a paired ratio between these two spaces of 3.6, 5.6 and 6.6 after 3, 10 and 20 min incubation, respectively.

(iv) *Generation of unesterified L-[1-^{14}C]glucose from β-L-[1-^{14}C]glucose pentaacetate:* the precise quantification of unesterified L-[1-^{12}C]glucose generation in islets exposed for 3 min to β-L-[1-^{14}C]glucose pentaacetate (1.7 mM) was hampered by the presence of radioactive molecules, presumably partially esterified L-[1-^{14}C]glucose (e.g., L-[1-^{14}C]glucose monoacetate), eluting together or in the vicinity of the unesterified hexose; nevertheless, the experimental results clearly indicated that the islets obviously and severely decreased the amount of β-L-[1-^{14}C]glucose pentaacetate and clearly increased that of its metabolites already after only 3 min incubation

(v) *Metabolism of β-L-glucose pentaacetate:* in sharp contrast to the results obtained with D-[1-^{14}C]glucose or D-[U-^{14}C]glucose, no significant production of $^{14}CO_2$ or ^{14}C-labelled acidic metabolites and amino acids could be detected in islets exposed to either L-[1-^{14}C]glucose or β-L-[1-^{14}C]glucose pentaacetate, all nutrients being tested at a 1.7 mM concentration. Both $^{14}CO_2$ and ^{14}C-labelled acidic metabolites were, however, produced by islets exposed to β-L-glucose penta-[1-^{14}C]acetate (also 1.7 mM) used to characterize the metabolic fate of the acetyl moiety of β-L-glucose pentaacetate.

(vi) *Interference of β-L-glucose pentaacetate with the catabolism of other nutrients:* the effects of β-L-glucose pentaacetate on the metabolism of islet endogenous nutrients was examined in islets prelabelled with either L-[U-^{14}C]glutamine or [U-^{14}C]palmitate. The ester failed to affect L-[U^{14}C]glutamine metabolism and slightly decreased the $^{14}CO_2$ output to ^{14}C content ratio of islets preincubated for 120 min in the presence of [U-^{14}C]palmitate. β-L-glucose pentaacetate (1.7 mM) failed to affect the oxidation of exogenous D-[U-^{14}C]glucose (8.3 mM), and decreased both the conversion of D-[5-^{3}H] glucose to ^{3}HOH and that of D-[U-^{14}C]glucose to either radioactive acidic metabolites or amino acids

(vii) *Biosynthetic data:* β-L-Glucose pentaacetate (1.7 mM) caused a modest but significant decrease of the incorporation of L-[4-^{3}H]phenylalanine into TCA-precipitable material in islets incubated for 90 min in the presence of either D-glucose (4.2 mM) or L-leucine (10.0 mM)

(viii) *Secretory data*: β-L-Glucose pentaacetate (1.7 mM) only exerts an obvious stimulation of insulin release from islets incubated in the presence of a suitable nutrient, e.g., D-glucose (7.0 mM), L-leucine (10.0 mM) or succinic acid dimethyl ester (10.0 mM). The insulinotropic action of β-L-glucose pentaacetate (1.7 mM) in the presence of L-leucine (5.0 mM) was also documented in a reverse hemolytic plaque assay of insulin secretion from isolated β-cells. β-L-Glucose pentaacetate appeared less potent than α-D-glucose pentaacetate in terms of augmenting the secretory response to D-glucose or L-leucine and even opposed the insulinotropic action of α-D-glucose pentaacetate.

(ix) *Adenylate cyclase activity and cAMP formation*: The pentaacetate ester of β-L-glucose (1.7 mM) failed to affect both basal adenylate cyclase activity in a membrane-enriched islet subcellular fraction and cyclic AMP production by islets incubated for 60 min in the presence of L-leucine (10.0 mM) together with 1.0 mM isobutylmethylxanthine

(x) *Phosphoinositide hydrolysis*: β-L-Glucose pentaacetate (1.7 mM) had little effect upon the production of tritiated inositol phosphate by islet preincubated for 180 min at 5.0 mM D-glucose in the presence of *myo*-[2-^{3}H]inositol and then incubated for 30 min in the presence of 10.0 mM LiCl and 10.0 mM L-leucine.

(xi) *Intracellular pH*: β-L-Glucose pentaacetate (1.7 mM) had no obvious effect upon the intracellular pH of dispersed islet cells exposed to 8.3 mM D-glucose.

(xii) *Cationic data*: β-L-Glucose pentaacetate (1.7 mM) failed to affect or tended to decrease the net uptake of ^{45}Ca by islets incubated in the absence of any exogenous nutrient or in the presence of either D-glucose (8.3 mM) or L-leucine (10.0 mM), such a difference only achieving statistical significance in the presence of the amino acid. In the absence of any exogenous nutrient, β-L-glucose caused a rapid decrease in ^{86}Rb fractional outflow rate from prelabelled islets. This rapid, sustained, and rapidly reversible decrease in K^+ conductance was not associated with any obvious change in ^{45}Ca fractional outflow rate. Nevertheless, the administration of β-L-glucose pentaacetate caused a rapid increase in insulin output, the secretory rate being already significantly higher than the paired basal value within 2 min exposure to the ester. In the presence of the dimethyl ester of succinic acid (10 mM), the administration of β-L-glucose pentaacetate to prelabelled and perifused islets caused a minor increase in ^{45}Ca fractional outflow rate and a rapid increase in insulin output.

(xiii) *Electrophysiological data*: In the presence of 4.0 mM D-glucose or 10.0 mM L-leucine, the administration of β-L-glucose pentaacetate (1.7 mM) to rat B-cells induced within 1–2 min depolarization of the plasma membrane and induced electrical activity. Comparable results were observed in mouse pancreatic islets, the induction of electrical activity being, on occasion, associated with oscillations of the cytosolic Ca^{2+} concentration [58].

A further study revealed that in isolated perfused rat pancreases exposed to 10.0 mM L-leucine, β-L-glucose pentaacetate (1.7 mM) stimulated both insulin and somatostatin release and provoked an initial and short-lived stimulation of glucagon secretion, in sharp contrast to the immediate inhibitory action of unesterified D-glucose upon glucagon output. A direct effect of the ester itself, by some as-of-yet unidentified coupling process, was speculated to account for the stimulation of insulin and somatostatin release by β-L-glucose pentaacetate and for the initial enhancement of glucagon secretion by the same ester [59].

The results of these experiments were interpreted to indicate that the insulinotropic action of β-L-glucose pentaacetate is not attributable to any nutritional value of the ester but, instead, appears to result from a direct effect of the ester itself on a yet unidentified receptor system, resulting in a decrease of K^+ conductance, plasma membrane depolarization, and induction of electrical activity [56, 59].

The insulin secretory response to β-L-glucose pentaacetate (1.7 mM) of islets incubated in the presence of L-leucine (10.0 mM) was not adversely affected when the islets were preincubated for 90 min at 37 °C in the presence of cholera toxin (5 μg/ml) or obtained from animals injected intraperitoneally 72 h before sacrifice with pertussis toxin (2.8 μg/rat); these findings suggest that G-proteins sensitive to either cholera or pertussis toxin are not involved in the modality by which β-L-glucose pentaacetate stimulates insulin secretion [60].

Selected monosaccharide pentaacetate esters (1.7 mM) were eventually found to display a bitter taste. Scores of 4, 3, and 2 arbitrary units were used, respectively, for very bitter, bitter, or slightly bitter solutions, a score of 1 corresponding to doubtful answers and a null score to solutions devoid of bitter state. Both α-D-glucose pentaacetate (1.98 ± 0.31 arbitrary units; n = 24) and β-L-glucose pentaacetate (2.87 ± 0.36; n = 16) displayed a bitter taste, the latter two mean scores being not significantly different from one another and yielding an overall mean value of 2.34 ± 0.24 arbitrary units (n = 40). Likewise, the mean scores found with β-L-glucose pentaacetate (2.00 ± 0.36; n = 16) and α-D-mannose pentaacetate (3.12 ± 0.31; n = 16) failed to differ significantly from that found for the anomers of D-glucose pentaacetate. Neither α-D-galactose pentaacetate nor β-D-galactose pentaacetate yielded a mean score significantly higher than the limit of detection for bitter state (i.e., 1.0 arbitrary unit). None of the esters tested in this study were identified as sweet by any of the subjects who tasted them, in sharp contrast with the results recorded for the sweet taste of D-glucose (0.2 mM), L-glucose (0.2 mM), D-galactose (0.22 mM), and D-mannoheptulose (0.22 mM). It was proposed that the interaction between these esters and a protein involved in the recognition of bitter taste may participate in their insulinotropic action [61].

In a review article entitled "The riddle of L-glucose pentaacetate insulinotropic action," it was underlined that the postulated interaction of L-glucose pentaacetate with a yet unidentified receptor leading to membrane depolarization, induction of electrical activity, and increase in cytosolic concentration of ionized Ca^{2+} displays analogies with the identification of bitter compounds by taste buds. Purified islet B-cells were proposed to contain the α-gustducin G-protein involved in the perception of bitter taste by taste buds. The effects of β-L-glucose pentaacetate upon glucagon and somatostatin secretion by the isolated perfused pancreas were considered to be also compatible with such a hypothesis. It was even proposed that L-glucose pentaacetate anomers could conceivably be used as novel insulinotropic tools in the treatment of noninsulin-dependent diabetes mellitus [54].

Further support in such a perspective was provided by the finding that the intravenous injection of β-L-glucose pentaacetate (8.8 mmol/g body wt) into fed anaesthetized rats caused, over 30 min, a biphasic increase in plasma insulin concentration, which could not be attributed to the modest rise in plasma D-glucose concentration recorded in these experiments [62]. It was also documented that β-L-glucose pentaacetate can be safely used in human subjects in further work aiming at the investigation of its insulinotropic and metabolic effects [63].

Five years later, denatonium, one of the most bitter-tasting substances known, was found to stimulate insulin release in both clonal HIT-T15 beta cells and rat pancreatic islets. This insulinotropic action, documented in the presence of 8.3 mM D-glucose, was abolished in the absence of extracellular Ca^{2+} or in the presence of the Ca^{2+}-channel blocker nitrendipine and inhibited by the α_2-adrenergic agonist clonidine. Furthermore, it could not be attributed to any distinct effect on voltage-gated calcium channels or cellular cyclic AMP levels, and no evidence was found to suggest activation by denatonium of either gustducin or transducing in the beta cells. The insulinotropic action of this bitter compound was eventually ascribed to its interaction with ATP-responsive K^+ channels, leading to the depolarization of beta cells and resulting increase in Ca^{2+} influx [64].

1.10 Effects of Artificial Sweeteners on Insulin Release

In the light of the just mentioned findings concerning the stimulation of insulin secretion by the β-anomer of L-glucose pentaacetate, a further study reported in 1998 aimed at re-evaluating the possible effect of both bitter and nonbitter artificial sweeteners on insulin release and cationic fluxes in isolated rat pancreatic islets [65]. Sodium saccharine (1.0–10.0 mM), sodium cyclamate (5.0–10.0 mM), stevioside (1.0 mM), and acesulfame-K (1.0–15.0 mM), all of which display a bitter taste, augmented insulin release from islets incubated in the presence of 7.0 mM D-glucose. In contrast, aspartame (1.0–10.0 mM), which is devoid of bitter taste, failed to affect insulin secretion. A positive response to acesulfame-K was still observed when the extracellular K^+ concentration was adjusted to the same value as that in control media. Saccharin (10.0 mM) and cyclamate (also 10.0 mM) even significantly augmented insulin output from islets incubated in the absence of exogenous D-glucose and, to a lesser relative extent, from islets exposed to 20.0 mM D-glucose. The enhancing action of both sodium saccharin and sodium cyclamate on insulin release from islets incubated for 90 min in the presence of 7.0 mM D-glucose was concentration-related in the 1.0–5.0 to 10.0 mM range. Niki and Niki had also drawn attention in 1994 to the fact that acetosulfame-Na (10.0 mM) augments insulin release evoked by 10.0 mM D-glucose [66]. No major changes in ^{86}Rb and ^{45}Ca outflow from prelabelled perifused islets could be attributed to the saccharin, cyclamic, or acetosulfame anions. It was proposed, therefore, that the insulinotropic action of some artificial sweeteners, like that of selected hexose pentaacetate esters, may imply G-protein-coupled receptors similar to those operative in the recognition of bitter compounds by tasted buds [65].

1.11 Glucotoxicity: The Anomeric Malaise

The information so far provided on the fate of D-glucose anomers in islet cells is far from covering all findings so far collected in this issue. For instance, attention was also paid to the low mutarotase activity in normal and tumoral pancreatic islet cells [67], on the anomeric specificity of glucose metabolism in the pentose cycle both in rat pancreatic islets and tumoral insulin-producing cells [68], on the anomeric specificity of hexokinase and glucokinase in normal pancreatic islets or insulin-producing tumoral cells [69], on the reciprocal influence of glucose anomers upon their respective phosphorylation by hexokinase in homogenates of tumoral islet cells [70], and on the anomeric dissociation between glucokinase activity and glycolysis in pancreatic islets [71], to cite only a few examples. This issue concerning the anomeric specificity of hexose metabolism in pancreatic islets was also the object of three reviews [72–74].

In my opinion, one of the most essential information concerning the effects of D-glucose anomers relates to the anomeric malaise phenomenon, one of the two most obvious aspects of the process of B-cell glucotoxicity.

It was generally admitted that in type 2 diabetes, the secretory response of beta cells to D-glucose is more severely affected than that evoked by other nutrient or nonnutrient secretagogues. Five potential candidates for perturbation of glucose metabolism in beta cells were considered in the framework of a concept presented as a G quintet [75]. The five candidates were **G**lut2 underexpression, **g**lucokinase mutation, **g**lucose-6-phosphatase hyperactivity, **g**lycerophosphate dehydrogenase deficiency, the latter FAD-linked mitochondrial enzyme playing a key role in the shuttle for the transfer of reducing equivalents from the cytosol into mitochondria, and **g**lycogen accumulation in beta cells as observed in situations of sustained hyperglycemia.

An anomaly of glucose transport in insulin-producing cells was first observed in tumoral islet cells of the RINm5F line [76–78]. Underexpression of Glut2 was then observed not only in diabetic subjects but also in four animal models of type 2 diabetes [79–83]. A mutation of the glucokinase gene was first identified in 1992–1993 in subjects with maturity-onset diabetes of the young (MODY) [84, 85]. An excessive activity of islet glucose-6-phosphatase was observed in *ob/ob* mice [86]. A decreased activity of mitochondrial glycerophosphate dehydrogenase was documented in islets from *db/db* mice, GK rats, and *fa/fa* rats [87–89]. Last, the accumulation of glycogen in insulin-producing cells represents a key feature of the phenomenon of insular glucotoxicity. The latter concept refers to the functional perturbations of beta cells provoked by sustained hyperglycemia. The two phenomenological aspects of this glucotoxicity process consist in a paradoxical and transient decrease in insulin output in response to a rapid increase in the extracellular concentration of D-glucose and a perturbation of the anomeric specificity of the insulin secretory response to the hexose.

Both anomalies were proposed to be attributable to glycogen accumulation in islet B-cells.

The paradoxical and transient inhibition of insulin release in response to a rise in extracellular D-glucose concentration was first observed in many noninsulin-dependent diabetic subjects after intravenous administration of D-glucose [90]. The latter anomaly was abolished in the same subjects when they were again examined after 20 h of glycemia normalization by insulin infusion [90]. The restauration of a positive insulin secretory response to glucose persisted in these subjects even when the second hyperglycemic test was performed 30 or 60 min after the end of insulin infusion followed by intravenous administration of glucose in order to bring the glycemia at the same high level as that recorded before the first hyperglycemic test conducted the day before [91]. A paradoxical response to either an increase or decrease in extracellular D-glucose concentration was also observed in the perfused pancreas of rats infused for 48 h with a hypertonic D-glucose solution [92, 93].

The perturbation of the anomeric specificity of the beta cell secretory response to D-glucose, referred to as an anomeric malaise [94], was also first observed in a study comparing the rapid changes in plasma insulin concentration following the intravenous administration of α- and β-D-glucose to either normal subjects or noninsulin-dependent diabetic patients. In seven normal subjects, the α/β paired ratio for plasma insulin concentration (expressed relative to the reference value measured prior to glucose injection) averaged 160, 129, and 115%, respectively 2, 4, and 6 min after administration of the D-glucose anomers. In eight diabetic subjects, the insulin response to glucose was too weak to allow characterization of its anomeric specificity. In the last five other diabetic subjects, a preferential response to α-D-glucose was observed in three cases, but not so in the last two cases. The severity of diabetes as judged from the plasma glucose and insulin concentrations was more pronounced in the two diabetic subjects with an α/β ratio for plasma insulin concentration not exceeding 88 ± 5%, three to six min after injection of D-glucose than in the three diabetic subjects with a mean value for the same ratio of 151 ± 11% over the same period. The eight subjects in whom the anomeric specificity of the B-cell secretory response could not be established were eventually identified as the most severely diabetic patients [95].

Complementary information on the anomeric malaise was then obtained in different animal models. For instance, in BioBreeding diabetic rats about 125 days old with a plasma glucose concentration close to 25 mM, no difference between secretory response to α- or β-D-glucose was observed in the perfused pancreas, while in control rats of the same age with a plasma glucose concentration close to only 8 mM, the response to β-D-glucose only represented about the two-thirds of that found with α-D-glucose. The α/β ratio for the increment in insulin output averaged no more than 60% in the diabetic rats, as distinct from 235% in the control normoglycemic rats [96]. Incidentally, when both the control rats and those with a high incidence of diabetes were examined at a younger age (60–65 days) and were all normoglycemic, the insulinotropic capacity of α-D-glucose largely exceeded that of β-D-glucose [74]. These observations suggested that the anomeric malaise in BB rats resulted from sustained hyperglycemia. The latter hypothesis was compatible with the fact that the anomeric preference for α-D-glucose as an insulin secretagogue was also obvious in female Zucker rats (*fa/fa*), which, at variance with male Zucker rats, maintain a close-to-normal glycemia at the age (13–26 weeks) of the animals indeed used for the experiments in the isolated perfused pancreas [97]. However, a loss of the anomeric specificity of the insulin secretory response to D-glucose was also observed in adult rats which had received an intraperitoneal administration of streptozotocin during the neonatal period [98]. Likewise, in rabbits examined 4–6 weeks after ligation of the pancreatic duct, the preference for α-D-glucose, otherwise observed in control rabbits, was no more obvious. As a matter of fact, the insulin secretory response to β-D-glucose remained unchanged in control rabbits and those with ligated pancreatic duct, whether the latter animals remained euglycemic or became frankly hyperglycemic after surgery, with an overall mean value for the increment in insulin output relative to paired basal value of 55 ± 9%. In contrast, the secretory response to α-D-glucose, expressed in the same manner, progressively decreased from 84 ± 19% in the control rabbits to 65 ± 21% and 31 ± 2% in the close-to-normoglycemic and hyperglycemic duct-ligated rabbits, respectively [99]. These observations were thus compatible with the concept that the relative severity of the anomeric malaise was linked to the extent of glucose intolerance, as already suggested by the experiments conducted in diabetic human subjects and BB rats. The validity of the latter concept was further documented in rats made modestly and briefly hyperglycemic by repeated oral administration of diazoxide and glucose during a period of 48 h before pancreas perfusion. Before presurgical anesthesia, the plasma concentration of glucose averaged in these rats 15.4 ± 2.1 mM as distinct from 7.3 ± 0.6 mM in the control rats. The α/β ratio for the increment in insulin output provoked by the D-glucose anomers represented in the rats receiving diazoxide no more than half the value found in control animals [100]. Taken as a whole, these findings suggested that hyperglycemia provokes, as a function of its duration and severity, first an attenuation and then a suppression, if not inversion, of the anomeric preference of insulin release for α-D-glucose.

Two further studies conducted, on one hand, in rats which underwent partial pancreatectomy a few months prior to perfusion of their pancreas and, on the other hand, in rats fasted for 48 h prior to such a pancreatic perfusion documented the absence of any significant alteration of the anomeric specificity of glucose stimulated insulin secretion [101, 102]. The experiments thus indicated that the anomeric malaise, taken as a manifestation of glucotoxicity, was not attributable to either an hyperactive insulin secretion by the B-cells, such as that imposed to rats from which 80–85% of the pancreas had been removed, or a decrease in the insulin secretory response to glucose, as prevailing in islets during starvation.

The biochemical determinants of the process of B-cell glucotoxicity were eventually considered. The nonenzymatic glycation of intracellular proteins was not considered as a likely explanation. For instance, the nonenzymatic glycation of phosphoglucose isomerase, an enzyme which plays an essential role in the anomeric specificity of the insulin secretory response to D-glucose, does not alter its specific activity, affinity for glucose 6-phosphate, isotopic discrimination, and anomeric specificity [103]. Likewise, the modest accumulation of sorbitol observed in the islets of rats infused for 48 h with a hypertonic solution of glucose is unlikely to account for the anomeric malaise, the insulin secretory response to glucose being comparable whether the islets maintained for 20 h at a high concentration of glucose were preincubated in the presence or absence of an aldose reductase inhibitor [104].

A set of arguments suggested, however, that the intracellular accumulation of glycogen in insulin-producing cells represents a major determinant of the phenomenon of insulin glucotoxicity. As already mentioned in this chapter, the interaction of phosphorylase *a* with D-glucose anomers may represent an essential aspect of glycogenolysis. The effects of D-glucose anomers upon the rate of glycogenolysis were also already considered [28].

Three series of experiments provided the essential information. First, the secretory behavior of insulin- and glucagon-producing cells was found to be perturbed in isolated perfused pancreases removed from rats infused with hypertonic solutions of glucose for 48 h. The anomalies included a high basal release of insulin and a paradoxical increase in insulin output and decrease in glucagon release in response to a fall in extracellular D-glucose concentration. Likewise, in isolated islets prepared from the glucose-infused rats, theophylline stimulates insulin release at a low ambient concentration of D-glucose, at variance with the situation found in islets removed from normal rats. These secretory perturbations coincided with an abnormal accumulation of glycogen in the B-cell [93].

Second, when glycogen-rich islets, first cultured for a few days in the presence of high concentrations of D-glucose (20–80 mM), were incubated for 6–10 min in the absence of glucose, the rate of glycogenolysis was grossly proportional to the islet glycogen content. Exogenous D-glucose (7–20 mM) inhibited glycogenolysis, the latter effect opposing the increase in glycolytic flux attributable to the utilization of exogenous glucose. Both the inhibitory effect of D-glucose on glycogenolysis and the utilization of the exogenous hexose tended to be higher with α- than with β-D-glucose [28].

In the last study, rats were perfused for 48–72 h with a hypertonic solution of glucose. The islets of these rats display a paradoxical and transient increase in insulin output when the extracellular concentration of D-glucose is decreased from 16.7 to 2.8 mM after 45 min of perifusion. The glycogen content of the islets initially averaged 37 picomoles of glucose residues per islet. The islets were then incubated for two successive periods of 10 min each: first in the presence of 16.7 mM D-glucose and then in the presence of only 2.8 mM exogenous D-glucose. The total output of lactic acid was comparable during these two successive incubations. However, when D-[U-^{14}C]glucose was present in the incubation medium, the lactic acid-specific radioactivity was almost four times lower during the second than first incubation, indicating a considerably increased contribution of glycogen-derived unlabeled glucose to overall glycolytic flux in response to the decrease in extracellular D-glucose concentration [105]. A quantitative model was proposed to account (i) for the progressive increase in basal insulin release, (ii) progressive shift in the anomeric specificity of glucose-stimulated insulin secretion, and (iii) paradoxical inhibition of insulin release by extracellular D-glucose as a function of the glycogen content of insulin-producing cells [106].

1.12 The Sweet Taste Receptor Tir3

In an impressive series of recent publications, Itaru Kojima and colleagues revealed, since 2009, the presence of a sweet taste receptor in pancreatic islet cells and documented its multimodal signaling process in these cells [107–116]. These crucial findings, which are detailed in the last chapter of the present book, allowed to propose that the fuel concept postulating that the stimulation of insulin release by nutrient secretagogues reflects their capacity to act as a fuel in pancreatic islet beta cells is not incompatible with a receptor hypothesis postulating the participation of cell-surface receptors in the recognition of selected nutrients as insulinotropic agents [117].

In the prolongation of Itaru Kojima investigations and with his help, the presence of the TIR3 receptor in human pancreatic islet B-cells [118] and the effects of both sucralose and lactisole upon the bioelectrical activity in insulin-producing cells [119, 120] were also recently investigated.

Prior work of I. Kojima and colleagues also concerned investigations on the calcium-sensing receptor [121], the expression of which in pancreatic B-cells and possible role in the regulation of insulin release had been first examined in 1999 [122–124].

1.13 Conclusion

Maybe, the major conclusion that could be drawn from the present historical perspective resides in the fact that despite not ignoring the limitation, if not incorrectness, of prior occasionally rather dogmatic opinions on the process of D-glucose identification by pancreatic islet B-cells as an insulin secretagogue, it seems now possible to reconcile the receptor and fuel hypotheses for such a process.

References

1. Cerasi E, Luft R. Diabetes mellitus – a disorder of cellular information transmissiion? In: Proceedings of the thirteenth Nobel symposium, Stockholm, John Wiley & Sons, Inc. New York;1970. P.349–354.
2. Randle PJ, Hales CN. Chapter 13: Insulin release mechanism. In Steiner DF, Freinkel N (eds) Handbook of physiology, section 7, vol 1: endocrine pancreas. Washington, DC: American Physiological Society; 1972. P. 219–35.
3. Matschinsky F. Chapter 11: Enzymes, metabolites, and cofactors involved in intermediatary metabolism of islets of Langerhans. In Steiner DF, Freinkel N (eds) Handbook of physiology, section 7, vol 1: endocrine pancreas. Washington, DC: American Physiological Society; 1972. P. 199–214.
4. Niki A, Niki H, Miwa I, Okuda J. Insulin secretion by anomers of D-glucose. Science. 1974;186:150–1.
5. Grodsky GM, Fanska R, West L, Manning M. Anomeric specificity of glucose-stimulated insulin release: evidence for a Glucoreceptor? Science. 1974;186:536–8.
6. Rossini AA, Soeldner JS, Hiebert JM, Weir GC, Gleason RE. The effect of glucose Anomers upon insulin and glucagon secretion. Diabetologia. 1974a;10:795–9.
7. Matschinsky FM, Pagliara AS, Hover BA, Haymond MW, Stillings SN. Differential effects of alpha- and beta-D-glucose on insulin and glucagon secretion from the isolated perfused rat pancreas. Diabetes. 1975;24:369–72.
8. Idahl L-Å, Sehlin J, Täljedal I-B. Metabolic and insulin-releasing activites of D-glucose Anomers. Nature. 1975;254:75–7.
9. Grodsky GM, Fanska R, Lundquist I. Interrelationships between α and β anomers of glucose affecting both insulin and glucagon secretion in the perfused rat pancreas. Endocrinology. 1975;97:573–580.
10. Pierce M, Freinkel N. Anomeric specificity for the rapid transient efflux of phosphate ions from pancreatic islets during secretory stimulation with glucose. Biochem Biophys Res Comm. 1975;63:870–4.
11. Grill V, Cerasi E. Glucose-induced cyclic AMP accumulation in rat islets of Langerhans: preferential effect of the alpha Anomer. FEBS Lett. 1975;54:80–3.

12. A.A. Rossini, M Berger, J Shadden, GF Cahill. Beta cell protection to alloxan necrosis by anomers of D-glucose. Science 183: 424, 1974b.
13. Rossini AA, Arcangeli MA, Cahill GF. Studies of alloxan toxicity on the beta cell. Diabetes. 1975;24:516–22.
14. Miwa JO, Niki H, Niki A. Uptake of radioactive D-glucose anomers by pancreatic islets. J Biochem. 1975;78:1309–11.
15. Malaisse WJ, Sener A, Koser M, Herchuelz A. Identification of the A-stereospecific Glucosensor in the pancreatic B-cell. FEBS Lett. 1976a;65:131–4.
16. Malaisse WJ, Sener A, Koser M, Herchuelz A. The stimulus-secretion coupling of glucose-induced insulin release. XXIV. The metabolism of α- and ß-D-glucose in isolated islets. J Biol Chem. 1976b;251: 5936–43.
17. Malaisse WJ. Insulin release: the glucoreceptor myth. Med Sci Res. 1987;15:65–7.
18. Meglasson MD, Matschinsky FM. New perspectives on pancreatic islet glucokinase. Am J Phys. 1984;246:E1–E13.
19. Garfinkel D, Garfinkel L, Meglasson MD, Matschinsky FM. Computer modeling identifies glucokinase as glucose sensor of pancreatic beta-cells. Am J Phys. 1984;247:R527–36.
20. R.P. Robertson, D. Porte Jr. The glucose receptor. A defective mechanism in diabetes mellitus distinct from the beta adrenergic receptor. J Clin Invest 52: 870–876, 1973.
21. Levy J, Herchuelz A, Sener A, Malaisse WJ. The stimulus-secretion coupling of glucose-induced insulin release. XX. Fasting: a model for altered glucose recognition by the b-cell. Metabolism. 1976;25:583–91.
22. Garvey WT, Olefsky JM, Griffin J, Hamman RF, Kolterman OG. The effect of insulin treatment on insulin secretion and insulin action in type II diabetes mellitus. Diabetes. 1985;34:222–34.
23. Sener A, Malaisse WJ. L-leucine and a nonmetabolized analogue activate pancreatic islet glutamate dehydrogenase. Nature. 1980;288:187–9.
24. Sener A, Malaisse-Lagae F, Malaisse WJ. Stimulation of islet metabolism and insulin release by a nonmetabolizable amino acid. Proc Natl Acad Sci U S A. 1981;78:5460–4.
25. Malaisse WJ, Sener A, Malaisse-Lagae F. Insulin release: reconciliation of the receptor and metabolic hypotheses. Nutrient receptors in islet cells. Mol Cell Biochem. 1981;37:157–65.
26. Malaisse WJ, Sener A, Koser M, Ravazzola M, Malaisse-Lagae F. The stimulus-secretion coupling of glucose-induced insulin release. XXV. Insulin release due to glycogenolysis in glucose-deprived islets. Biochem J. 1977;164:447–54.
27. Bollen M, Malaisse-Lagae F, Malaisse W, Stalmans W. The interaction of phosphorylase *a* with D-glucose displays a-stereospecificity. Biochim Biophys Acta. 1990;1038:141–5.
28. Malaisse WJ, Marynissen G, Sener A. Possible role of glycogen accumulation in B-cell Glucotoxicity. Metabolism. 1992;41:814–9.
29. Malaisse WJ. Metabolic signalling of insulin secretion. Diab Rev. 1996;4:145–59.
30. Malaisse WJ, Malaisse-Lagae F, Davies DR, Van Schaftingen E. Presence of fructokinase in pancreatic islets. FEBS Lett. 1989;255:175–8.
31. Malaisse WJ, Malaisse-Lagae F, Davies DR, Vandercammen A, Van Schaftingen E. Regulation of glucokinase by a fructose-1-phosphate-sensitive protein in pancreatic islets. Eur J Biochem. 1990a;190:539–45.
32. Zähner D, Malaisse WJ. Kinetic behaviour of liver glucokinase in insulinopenic situations: effect of fructose 1-phosphate in fed and starved rats. Biochimie. 1990;72:715–8.
33. Zhang T-M, Giroix M-H, Sener A, Malaisse WJ. Glucokinase activation by its fructose phosphate-sensitive protein in the liver of rats with inherited or acquired non-insulin-dependent diabetes mellitus. Med Sci Res. 1995;23:395–6.
34. Malaisse WJ, Malaisse-Lagae F, Rasschaert J, Zähner D, Sener A, Davies DR, Van Schaftingen E. The fuel concept for insulin release: regulation of glucose phosphorylation in pancreatic islets. Biochem Soc Trans. 1990b;18:107–8.
35. Malaisse WJ, Malaisse-Lagae F, Sener A, Van Schaftingen E, Hers H-G. Is the glucose-induced stimulation of glycolysis in pancreatic islets attributable to activation of phosphofructokinase by fructose 2,6-bisphosphate? FEBS Lett. 1981b;125:217–9.

36. Malaisse WJ, Malaisse-Lagae F, Sener A. Presence of fructose-6-phosphate,2-kinase in pancreatic islets. FEBS Lett. 1981c;135:203–6.
37. Malaisse WJ, Malaisse-Lagae F, Sener A. Glucose-induced accumulation of fructose-2,6-bisphosphate in pancreatic islets. Diabetes. 1982a;31:90–3.
38. Sener FM-L, Malaisse WJ. Glucose-induced accumulation of glucose-1,6-bisphosphate in pancreatic islets: its possible role in the regulation of glycolysis. Biochem Biophys Res Commun. 1982;104:1033–40.
39. Malaisse WJ, Malaisse-Lagae F, Sener A. The glycolytic cascade in pancreatic islets. Diabetologia. 1982b;23:1–5.
40. Sener A, Van Schaftingen E, Van De Winkel M, Pipeleers DG, Malaisse-Lagae F, Malaisse WJ, Hers H-G. Effects of glucose and glucagon on the fructose 2,6-bisphosphate content of pancreatic islets and purified pancreatic B-cells. Biochem J. 1984;221:759–64.
41. Malaisse WJ. The possible link between glycolysis and insulin release in isolated islets. In Lindenlaub E (ed) Diabetes research today. Stuttgart: F.K. Schattauer Verlag; 1976. P. 191–206.
42. Malaisse WJ, Sener A, Herchuelz A, Hutton JC. Insulin release: the fuel hypothesis. Metabolism. 1979;28:373–86.
43. Malaisse WJ. Insulin release: the fuel concept. Diab Métab. 1983;9:313–20.
44. Malaisse WJ, Sener A, Levy J, Herchuelz A. The stimulus-secretion coupling of glucose-induced insulin release. XXII. Qualitative and quantitative aspects of glycolysis in isolated islets. Acta Diabet Lat. 1976c;13:202–15.
45. Sener A, Malaisse WJ. The metabolism of glucose in pancreatic islets. Diab Métab. 1978;4:127–33.
46. Malaisse WJ, Sener A. Glucokinase is not the pancreatic B-cell glucoreceptor. Diabetologia. 1985;28:520–7.
47. Malaisse WJ. Physiology and pathology of the pancreatic B-cell glucose-sensor device. The Morgagni prize lecture. In Crepaldi G, Tiengo A, Enzi G (eds) Diabetes, obesity and hyperlipidemias. Amsterdam: Elsevier Science Publishers; 1990. P. 3–22.
48. Malaisse WJ, Zhang Y, Jijakli H, Courtois P, Sener A. Metabolism of D-glucose anomers in rat pancreatic islets exposed to equilibrated D-glucose. Horm Metab Res. 2004a;36:281–5.
49. Malaisse WJ, Zhang Y, Jijakli H, Courtois P, Sener A. Enzyme-to-enzyme channelling in the early steps of glycolysis in rat pancreatic islets. Int J Biochem Cell Biol. 2004b;36:1510–20.
50. Malaisse WJ. Monosaccharide esters: new tools in biomedical research. Mol Gen Metab. 1998a;65:129–42.
51. Malaisse WJ. The esters of carboxylic nutrients as Insulinotropic tools in non-insulin-dependent diabetes mellitus. Gen Pharmacol. 1995;26:1133–41.
52. Malaisse WJ, Rasschaert J, Villanueva-Penacarrillo ML, Valverde I. Respiratory, ionic and functional effects of succinate esters in pancreatic islets. Am J Phys. 1993a;264:E428–33.
53. Malaisse WJ, Sener A. Metabolic effects and fate of succinate esters in pancreatic islets. Am J Phys. 1993;264:E434–40.
54. Malaisse WJ. The riddle of L-glucose pentaacetate insulinotropic action. Int J Mol Med. 1998b;2:383–8.
55. Malaisse WJ. Insulinotropic action of monosaccharide esters: therapeutic perspectives. Diabetologia. 1999;42:286–91.
56. Malaisse WJ, Best LC, Herchuelz A, Hiriart M, Jijakli H, Kadiata MM, Larrieta-Carasco E, Laghmich A, Louchami K, Mercan D, Olivares E, Sánchez-Soto C, Scruel O, Sener A, Valverde I, Villanueva-Peñacarrillo ML, Viñambres C, Zawalich WS. Insulinotropic action of ß-L-glucose Pentaacetate. Am J Physiol. 1998a;275:E993–El006.
57. Olivares E, Kadiata MM, Malaisse WJ. Isomeric and anomeric specificity of hexose pentaacetate hydrolysis in rat pancreatic islet homogenates. Med Sci Res. 1998;26:489–90.
58. Pomares R, Ropero AB, Sanchez-Andres JV, Nadal A, Soria B, Malaisse WJ. Effect of hexose pentaacetates on electrical activity and cytosolic Ca2+ in mouse pancreatic islets. Int J Mol Med. 1999;3:15–20.

59. Leclercq-Meyer V, Malaisse WJ. Dual mode of action of glucose Pentaacetates on hormonal secretion from the isolated perfused rat pancreas. Am J Phys. 1998;275:E610–7.
60. Malaisse WJ, Ladriere L. Resistance of the insulinotropic action of α-D-glucose and ß-L-glucose pentaacetates to cholera and pertussis toxins. Biochem Mol Biol Int; 1998;44: 1127–1131.
61. Malaisse WJ, Malaisse-Lagae F. Bitter taste of monosaccharide pentaacetate esters. Biochem Mol Biol Int. 1997;43:1367–71.
62. García-Martínez JA, Cancelas J, Villanueva-Peñacarrillo ML, Valverde I, Malaisse WJ. Stimulation of insulin release by ß-L-glucose pentaacetate in anaesthetized rats. Med Sci Res. 1999;27:467–9.
63. Malaisse WJ, Ladriere L, Malaisse-Lagae F, Mercan D, Bastin G, Kadiata MM, Pauwels O, Benammar L. Innocuity of α-D-glucose pentaacetate and ß-L-glucose pentaacetate in healthy subjects. Diab Res. 1999a;34: 73–78.
64. Straub SO, Mulvancy-Musa J, Yajima H, Weiland GA, Sharp GWG. Stimulation of insulin secretion by denatonium, one of the most bitter-tasting substance known. Diabetes. 2003;52:356–64.
65. Malaisse WJ, Vanonderbergen A, Louchami K, Jijakli H, Malaisse-Lagae F. Effects of artificial sweeteners on insulin release and cationic fluxes in rat pancreatic islets. Cell Signal. 1998b;10:727–33.
66. Niki A, Niki H. Similarities between the hexose-recognizing mechanisms of the pancreatic B-cell and the gustatory cell. In: Flatt P, Lenzen S (eds) Frontiers of insulin secretion and pancreatic B-cell research. London: Smith-Gordon; 1994. P. 83–90.
67. Rasschaert J, Zähner D, Malaisse WJ. Low mutarotase activity in normal and tumoral pancreatic islet cells. Med Sci Res. 1987;15: 621–622.
68. Malaisse WJ, Giroix M-H, Sener A. Anomeric specificity of glucose metabolism in the pentose cycle. J Biol Chem. 1985;260:14630–2.
69. Sener M-HG, Dufrane SP, Malaisse WJ. Anomeric specificity of hexokinase and glucokinase activities in liver and insulin-producing cells. Biochem J. 1985a;230:345–51.
70. Giroix M-H, Sener A, Malaisse WJ. Reciprocal influence of glucose anomers upon their respective phosphorylation by hexokinase. Biol Chem Hoppe Seyler. 1986a;367:47–51.
71. Sener A, Giroix M-H, Leclercq-Meyer V, Marchand J, Malaisse WJ. Anomeric dissociation between glucokinase activity and glycolysis in pancreatic islets. Biochem Int. 1985b;11:77–84.
72. W.J. Malaisse, M. Deleers, F. Malaisse-Lagae, A. Sener. Anomeric specificity of hexose metabolism in pancreatic islets. In: Mngola EN (ed) Diabetes 1982. Excerpta Medica I.C.S. 1983a;600: 345–51.
73. Malaisse WJ, Malaisse-Lagae F, Sener A. Anomeric specificity of hexose metabolism in pancreatic islets. Physiol Rev. 1983b;63:321–7.
74. Malaisse WJ, Malaisse-Lagae F, Leclercq-Meyer V. Anomeric specificity of hexose metabolism in health and disease. Netherlands J Med. 1988;33:247–56.
75. Malaisse WJ. Alteration of pancreatic B-cell D-glucose metabolism in type 2 diabetes: the G quintet. Endocrinologia. 1993;40:309–13.
76. Malaisse WJ, Giroix M-H, Malaisse-Lagae F, Sener A. 3-O-methyl-D-glucose transport in tumoral insulin-producing cells. Amer J Physiol. 1986;251:C841–6.
77. Giroix M-H, Sener A, Malaisse WJ. D-glucose transport and concentration in tumoral insulin-producing cells. Am J Physiol. 1986b;251:C847–51.
78. Sener M, Giroix A-H, Malaisse WJ. Impaired uptake of D-glucose by tumoral insulin-producing cells. Biochem Internat. 1986;12:913–9.
79. Unger RH. Diabetic hyperglycemia: link to impaired glucose transport in pancreatic B-cells. Science. 1991;251:I.200–5.
80. Thorens B, Wu Y-J, Leahy JL, Weir GC. The loss of glut 2 expression by glucose unresponsive B-cells of db/db mice is reversible and is induced by the diabetic environment. J Clin Invest. 1992;90:77–85.

81. Thorens B, Weir GC, Leahy JL, Lodish HF. Reduced expression of the liver/B-cell glucose transporter isoform in glucose-insensitive pancreatic B-cells of diabetic rats. Proc Natl Acad Sci USA. 1990;87:6.492–6.
82. Ohneda M, Johnson JH, Inman LR, Chen L, Suzuki K-I, Goto Y, et al. Glut 2 expression and function in B-cells of Gk rats with Niddm. Dissociation between reductions in glucose transport and glucose-stimulated insulin secretion. Diabetes. 1993;42:1065–72.
83. Giroix M-H, Rasschaert J, Sener A, Leclercq-Meyer V, Bailbe D, Portha B, Malaisse WJ. Study of hexose transport, glycerol phosphate shuttle and Krebs cycle in islets of adult rats injected with Streptozotocin during the neonatal period. Mol Cell Endocrinol. 1992;83:95–104.
84. Vionnet N, Stoffel M, Takeda J, Yasuda K, Bell GI, Zouali H, et al. Nonsense mutation in the glucokinase gene causes early-onset non-insulin-dependent diabetes mellitus. Nature (Lond). 1992;356:721–2.
85. Shimada F, Makino H, Hashimoto N, Taira M, Seino S, Bell GI, et al. Type 2 (non-insulin-dependent) diabetes mellitus associated with a mutation of the glucokinase gene in a Japanese family. Diabetologia. 1993;36:433–7.
86. Khan A, Chandramouli V, Östenson G-C, Ahren B, Schumann WE, Löw H, et al. Evidence for the presence of glucose cycling in pancreatic islets of the Ob/Ob mouse. J Biol Chem. 1989;246:9.732–3.
87. Östenson G-C, Abdel-Halim SM, Rasschaert J, Malaisse-Lagae F, Meuris S, Sener A, et al. Deficient activity of FAD-linked glycerophosphate dehydrogenase in islets of Gk rats. Diabetologia. 1993;36:722–6.
88. Sener A, Herberg L, Malaisse WJ. FAD-linked glycerophosphate dehydrogenase deficiency in pancreatic islets of mice with hereditary diabetes. FEBS Lett. 1993;361:224–7.
89. Rasschaert J, Malaisse-Lagae F, Sener A, Leclercq-Meyer V, Herberg L, Malaisse WJ. Impaired FAD-glycerophosphate dehydrogenase activity in islet and liver homogenates of fa/fa rats. Mol Cell Biochem 1994;135:137–141.
90. Gomis R, Novials A, Coves MJ, Casamitjana R, Malaisse WJ. Suppression by insulin treatment of glucose-induced inhibition of insulin release in non-insulin-dependent diabetics. Diab Res Clin Pract. 1989;6:191–8.
91. Novials A, Gomis R, Coves MJ, Conget I, Malaisse WJ. Improvement by insulin treatment of the early phase of glucose-stimulated insulin release in non-insulin dependent diabetics explored after restoration of hyperglycemia. Med Sci Res. 1989;17:381–2.
92. Leahy JL, Cooper HE, Deal DA, Weir GC. Chronic hyperglycemia is associated with impaired glucose influence on insulin secretion. A study in normal rats using chronic in vivo glucose infusion. J Clin Invest. 1986;77:908–15.
93. Marynissen G, Leclercq-Meyer V, Sener A, Malaisse WJ. Perturbation of pancreatic islet function in glucose-infused rats. Metabolism. 1990;39:87–95.
94. Malaisse WJ. The anomeric malaise: a manifestation of B-cell glucotoxicity. Horm Metab Res. 1991a;23:307–11.
95. Rovira A, Garrotte FJ, Valverde I, Malaisse WJ. Anomeric specificity of glucose-induced insulin release in normal and diabetic subjects. Diab Res. 1987;5:119–24.
96. Leclercq-Meyer V, Marchand J, Malaisse WJ. Alteration of the insulin secretory response to D-glucose anomers in diabetic bb rats. Med Sci Res. 1987a;15:1535–6.
97. Leclercq-Meyer V, Marchand J, Malaisse WJ. Anomeric specificity of the insulin and glucagon secretory response to D-glucose in lean and obese Zucker rats. Pancreas. 1987b;2:645–52.
98. Niki A, Niki H, Niki I, Kunoh Y. Insulin release by glucose anomers in a rat model of non-insulin-dependent diabetes. Diabetologia. 1988;31:65–7.
99. Fichaux F, Marchand J, Yaylali B, Leclercq-Meyer V, Catala J, Malaisse WJ. Altered anomeric specificity of glucose-induced insulin release in rabbits with duct-ligated pancreas. Int J Pancreatol. 1991;8:151–67.

100. Leclercq-Meyer V, Marchand J, Malaisse WJ. Attenuated anomeric difference of glucose-induced insulin release in the perfused pancreas of diazoxide-treated rats. Horm Metab Res. 1991;23:257–61.
101. Leclercq-Meyer V, Malaisse-Lagae F, Coulic V, Akkan AG, Malaisse WJ. Preservation of the anomeric specificity of glucose-induced insulin release in partially Pancreatectomized rats. Diabetologia. 1992;35:505–9.
102. Leclercq-Meyer V, Marchand J, Malaisse WJ. Effect of starvation upon the anomeric specificity of glucose-induced insulin release. Diab Nutr Metab. 1993;6:129–34.
103. Zähner D, Liemans V, Malaisse-Lagae F, Malaisse WJ. Non-enzymatic glycation of Phosphoglucoisomerase. Diab Res. 1989;12:165–9.
104. Sener A, Malaisse WJ. Sorbitol generation and release in cultured islet B-cells. Med Sci Res. 1996;24:93–4.
105. Malaisse WJ, Maggetto C, Leclercq-Meyer V, Sener A. Interference of Glycogenolysis with glycolysis in pancreatic islets from glucose-infused rats. J Clin Invest. 1993b;91:432–6.
106. Malaisse WJ. Physiology of insulin secretion and its alteration in diabetes: the concept of Glucotoxicity. In: Andreani D, Gueriguian JL, Striker GE, editors. Diabetic complications: epidemiology and pathogenetic mechanisms. New York: Raven Press; 1991b. p. 3–23.
107. Nakagawa Y, Nagasawa M, Yamada S, Hara A, Mogami H, Nikolaev VO, Lohse MJ, Shigemura N, Ninomiya Y, Kojima I. Sweet taste receptor expressed in pancreatic ß-cells activates the calcium and cyclic amp signaling systems and stimulates insulin secretion. PLoS One. 2009;4:E5106.
108. Kojima I, Nakagawa Y. The role of the sweet taste receptor in Enteroendocrine cells and pancreatic B-cells. Diabetes Metab J. 2011;35:451–7.
109. Nakagawa Y, Nagasawa M, Mogami H, Lohse M, Ninomiya Y, Kojima I. Multimodal function of the sweet taste receptor expressed in pancreatic B-cells: generation of diverse patterns of intracellular signals by sweet agonists. Endocr J. 2013;60:1191–206.
110. Nakagawa Y, Ohtsu Y, Nagasawa M, Shibata H, Kojima I. Glucose promotes its own metabolism by acting on the cell-surface glucose-sensing receptor Tir3. Endocr J. 2014;61:119–31.
111. Li L, Ohtsu Y, Nakagawa Y, Masuda K, Kojima I. Sucralose, an activator of the glucose-sensing receptor, increases Atp by calcium-dependent and –independent mechanisms. Endocr J. 2016;63:715–25.
112. Kojima I, Nakagawa Y, Ohtsu Y, Hamano K, Medina J, Nagasawa M. Glucose activates the glucose-sensing receptor Tir3 and facilitates metabolism in pancreatic B-cells. J Diabetes Invest. 2015;6:256–63.
113. Medina A, Nakagawa Y, Ma J, Li L, Hamano K, Akimoto T, Ninomiya Y, Kojima I. Expression of the glucose-sensing receptor Tir3 in pancreatic islet: changes in the expression levels in various nutritional and metabolic states. Endocr J. 2014;61:797–805.
114. Kojima I, Medina J, Nakagawa Y. Role of glucose-sensing receptor in insulin secretion. Diabetes Obes Metab. 2017;19(Suppl 1):54–62.
115. Nakagawa Y, Nagasawa M, Medina J, Kojima I. Glucose evokes rapid Ca2+ and cyclic amp signals by activating the cell-surface glucose-sensing receptor in pancreatic B-cells. PLoS One. 2015;10:e0144053.
116. Hamano K, Nakagawa Y, Ohtsu Y, Li L, Medina J, Tanaka Y, Masuda K, Komatsu M, Kojima I. Lactisole inhibits the glucose-sensing receptor Tir 3 expressed in mouse pancreatic B-cells. J Endocrinol. 2015;226:57–66.
117. Malaisse WJ. Insulin release: the receptor hypothesis. Diabetologia. 2014;57:1287–90.
118. Malaisse WJ, Strange G, Kojima I. Presence of the sweet taste T1R3 receptor in human and rat pancreatic alpha cells. Abstract book autumn meeting Belg. Soc Physiol Pharmacol. 2014: P-02.
119. Sanchez-Andres JV, Malaisse WJ. Characterisation of the sweet taste receptor electrophysiology in pancreatic beta cells. In: Comm. 51st annual meeting of the European Association for the Study of Diabetes. 2015. Absract 429.

120. Sanchez-Andres JV, Malaisse WJ. Effects of lactisole on pancreatic B-cells electrophysiology. In: Abstract submitted for presentation at the 54th annual meeting of the European Association for the Study of Diabètes. 2018.
121. Medina J, Nakagawa Y, Nagasawa M, Fernandez A, Sakagushi K, Kitagushi T, Kojima I. Positive allosteric modulation of the calcium-sensing receptor by physiological concentrations of glucose. J Biol Chem. 2016;280:15238–46.
122. Rasschaert J, Malaisse WJ. Expression of the calcium-sensing receptor in pancreatic islet B-cells. Biochem Biophys Res Commun. 1999a;264:615–8.
123. Malaisse WJ, Louchami K, Laghmich A, Ladriere L, Morales M, Villanueva-Peñacarrillo ML, Valverde I, Rasschaert J. Possible participation of a islet B-cell calcium-sensing receptor in insulin release. Endocrine. 1999b;11:293–300.
124. Rasschaert J, Malaisse WJ. The G-protein coupled, extracellular Ca2+-sensing receptor: expression in pancreatic islet B-cells and possible role in the regulation of insulin release. Mol Gen Metab. 1999b;68:328–31.

Chapter 2
Cell-Surface Glucoreceptor Recognizing Anomers of Glucose in Pancreatic β-Cells

Yuko Nakagawa and Hatsumi Niki

Abstract Glucose is a major fuel stimulator of insulin secretion. When the effect of glucose anomers are compared in rat pancreatic islets, insulin secretion induced by the α-anomer is much more than that induced by the β-anomer. Similarly, insulin secretion induced by the α-mannose is higher than that induced by the β-anomer of mannose. It is well known that the α-anomer of hexose is stronger in activating the sweet taste receptor expressed in the taste buds of the tongue. Interestingly, inhibitors of the sweet taste receptor attenuate insulin secretion induced by glucose in pancreatic islets. These results raise an interesting possibility that glucose activates a cell-surface "glucoreceptor," which resembles the sweet taste receptor in the tongue, and exerts its action in pancreatic β-cells.

Keywords Glucose · Glucose anomer · Insulin secretion · Glucoreceptor · Sweet taste receptor

Glucose is an important energy source and also functions as a signaling molecule reflecting the energy state in the body. Therefore, the concentration of glucose is monitored strictly in vivo, and failure of the glucose-sensing system leads to disease conditions such as diabetes.

The pancreatic β-cell, one of the endocrine cells in the pancreas, secretes insulin in response to the elevation of the plasma glucose concentration. Since insulin is secreted only in pancreatic β-cells, these cells play a critical role in maintaining glucose homeostasis. It is well known that glucose is the most important stimulator of insulin secretion. Neurotransmitters such as acetylcholine and various hormones including glucagon-like peptide-1 (GLP-1) are also physiologically important regulators. However, these regulators cannot promote insulin secretion by themselves, nor can they stimulate secretion only in the presence of glucose. In contrast, glucose is an exceptional stimulator that can promote insulin secretion in the absence of any other agents. Hence, understanding the mechanism by which glucose promotes

Y. Nakagawa (✉) · H. Niki
Institute for Molecular and Cellular Regulation, Gunma University, Maebashi, Japan
e-mail: nakayuko@gunma-u.ac.jp; niki@hashioka.net

I. Kojima (ed.), *Glucose-sensing Receptor in Pancreatic Beta-cells*,
https://doi.org/10.1007/978-981-13-0002-8_2

insulin secretion is important, and many researchers have been investigating the issue. In the 1970s, there were two hypotheses as to the mechanism of action of glucose in pancreatic β-cells. One was the "metabolic hypothesis" in which glucose is metabolized intracellularly and its metabolic processes or metabolic products lead to stimulation of insulin secretion. The fact that blockade of the glucose metabolism suppresses insulin secretion is a major basis for this idea. The other is the "glucoreceptor hypothesis." This hypothesis is based on the idea that glucose exerts its action by binding to and activating the "glucoreceptor" localized in the plasma membrane. In the early days, it was simply an idea opposing the "metabolic hypothesis," but subsequently, a series of data supporting the glucoreceptor were reported. In this chapter, we would like to discuss the transition of the idea regarding the mechanism of action of glucose in pancreatic β-cells, featuring the glucoreceptor hypothesis.

Regarding the glucose-sensing system in pancreatic β-cells, the metabolic hypothesis and the glucoreceptor hypothesis existed in the 1960s, but the metabolic hypothesis was dominant. Meanwhile, Niki and colleagues in 1974 found that the α-anomer of D-glucose is more potent in stimulating insulin secretion than the β-anomer [1]. They isolated islets from Wistar rats by using collagenase and preincubated them for 30 min. For each islet batch, 300 μl each of a solution containing no D-glucose, pure α-D-glucose, pure β-D-glucose, or a mixture of two anomers (α: β, 36: 64) was added, and the islets were incubated for 5 min. At 37 °C, α-D-glucose is rapidly converted to β-D-glucose. The final concentration of each D-glucose was 2 mg/ml. Changes in the ratios of each anomer before and during incubation were determined by using β-D-glucose oxidase. After the incubation with each anomer, insulin was quantified by radioimmunoassay. The purity of each anomer before incubation was 98% or higher. After the 5 min incubation, the ratio of α-anomer to β-anomer was 42.4: 26.8. The amount of insulin in pancreatic islets cultured with each solution was as follows. The amount of insulin was 10.5 ± 1.1 μU/5 min in medium containing no D-glucose, 25.5 ± 2.3 μU/5 min in medium containing α-D-glucose, 18.5 ± 1.9 μU/5 min in medium containing β-D-glucose, and 20.4 ± 2.4 μU/5 min in medium containing two anomers. Indeed, α-D-glucose was approximately 1.5 times more potent than β-D-glucose. The amount of insulin secreted from pancreatic islets cultured with a mixture of two anomers was in the middle of amount of insulin from pancreatic islets cultured in a solution containing a single anomer. Considering that the conversion of α-D-glucose to β-D-glucose is rapid, the difference in the actual ability to secret insulin is more than these data demonstrate. Based on these findings, it is possible that a molecule(s) capable of discriminating α- and β-anomers is expressed in β-cells, and the action of glucose is mediated by the putative receptor molecule. Regarding the anomer, monosaccharides containing glucose have a cyclic structure in addition to the linear aldehyde type. When having a cyclic structure, the carbonyl carbon atom becomes a new asymmetric carbon atom, so that two diastereomers are generated due to the difference in arrangement of hydroxyl groups. These two diastereomers are called α- and β-anomers. The hydroxyl group of the carbonyl carbon atom and the substituent on the asymmetric carbon atom with the largest number are arranged in the trans, which is referred to as α-anomer, and the one arranged in the cis is called the

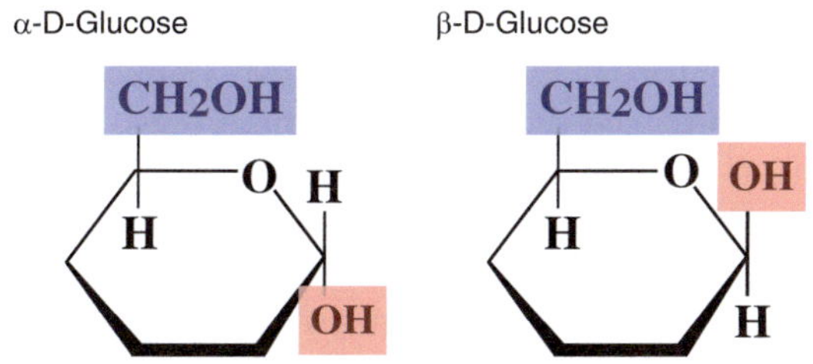

Fig. 2.1 Structure of α- and β-anomer of glucose. When monosaccharide has a cyclic structure, two diastereomers are generated because the carbonyl carbon atom is an asymmetric carbon atom. The hydroxyl group of the carbonyl carbon atom (red) and the substituent on the asymmetric carbon atom having the largest number (blue) are mutually trans isomers of α-anomer, and those of cis are β-anomer

β-anomer (Fig. 2.1). The mutarotation between anomers is rapid. Indeed 20% of the α-anomer converts to the β-anomer in only 2 min. After 5 min, about 40% of the α-anomer converts to the β-anomer and eventually reaches equilibrium. In the study by Niki and colleagues, they cultured pancreatic islets for 5 min with two anomers. Indeed, pancreatic β-cells discriminated two anomers and induced insulin secretion to a different extent. Given that the incubation was for only 5 min, their results suggest that the anomer is not identified in the process of metabolism, but rather it promotes insulin secretion by another mechanism independent of metabolism. Regarding this mechanism, involvement of two molecules should be considered. One is a glucose transporter. According to the previous reports, glucose uptake is not different between α and β anomers [2, 3]. It is also known that insulin secretion cannot be induced only by incorporation of glucose. Therefore, it seems quite likely that another candidate molecule, glucoreceptor, sensing the structural difference in glucose anomers controls insulin secretion. The glucoreceptor localized in the plasma membrane recognizes the difference in the conformation of each anomer, reflects this in signal transduction, and eventually induces insulin secretion. The ability of each glucose anomer to stimulate insulin secretion shown in their study suggests that the glucoreceptor plays an important role in glucose-induced insulin secretion. Accordingly, their results strongly support the glucoreceptor hypothesis. Subsequently, Grodsky and colleagues reported an article entitled "Anomeric specificity of glucose-stimulated insulin release: evidence for a glucoreceptor?" [4]. In perifusion experiments using pancreatic islets, various concentrations of α- and β-anomer were added and insulin secretion was compared. Compared to the β-anomer, α-anomer elicited insulin secretion at low concentrations, but no difference was found in the maximum stimulatory concentration. Indeed, pancreatic β-cells are more sensitive to α-anomer compared to β-anomer. In this report, the authors postulated that the glucoreceptor expressed in pancreatic β-cells detects differences in the conformation of the anomeric structure, and, when activated by glucose, the putative receptor evokes signals, which eventually induces insulin secretion. Collectively, these two studies provided support to the "glucoreceptor hypothesis" regarding the glucose-sensing system in pancreatic β-cells.

In addition to the above-mentioned reports, several lines of evidence supporting the "glucoreceptor hypothesis" have been reported. There is no difference in glucose transport between α- and β-anomers in various cells including β cells, or in some situations, the β-anomer predominates [2, 3]. Glucokinase expressed in pancreatic β cells does not discriminate anomeric differences [5, 6]. In addition, once glucose is phosphorylated, the rate of interconversion of anomers becomes 270 times higher, and at down-stream steps, both anomers of glucose are equally potent in β-cells [5]. Also, the production of H_2O by the two anomers is comparable [2, 7]. The concentration of glucose-6-phosphate is roughly comparable whether extracellular D-glucose is α-anomer or β-anomer, but rather, it is higher in the case of β-anomer [2, 7–9]. All of these results strongly support the "glucoreceptor hypothesis."

On the other hand, regarding the metabolic hypothesis, Malaisse and colleagues highlighted phosphoglucose isomerase, an enzyme that converts glucose 6-phosphate to fructose 6-phosphate. They showed that phosphoglucose isomerase is more efficient for α-anomer, and postulated that the difference in the action of α- and β-anomers depends on "metabolism" [8, 9].

In 1979, Niki and colleagues again stimulated β-cells by mannose anomers and compared insulin secretion induced by two anomers, focusing on phosphomannose isomerase. This enzyme more efficiently catalyzes the β-anomer. Accordingly, it is expected that if the structure of anomers is recognized by the enzyme, β-anomer is more potent in stimulating insulin secretion. Nevertheless, insulin secretion induced by the α-anomer of mannose is more similar to that induced by the β-anomer [10].

In 1983, Matchinsky reported that glucokinase, which was previously thought to be unable to distinguish α- and β-anomers, specifically recognizes α-anomer. They suggested that glucokinase perceives anomeric differences and causes a difference in insulin secretion. They also suggested that this glucokinase is a glucose sensor, since the Km value and the V_{max} value of glucokinase for glucose and mannose agree with the respective values of glucose metabolism in pancreatic β-cells. Their work provided support for the "metabolic hypothesis."

Meanwhile, results supporting the "glucoreceptor hypothesis" have been reported. Mannoheptulose is a metabolic inhibitor of glucokinase and has been shown to inhibit secretion of insulin. In 1988, Wolf and colleagues permeabilized plasma membrane of pancreatic β-cells with digitonin. Using these permeabilized cells, they showed that mannoheptulose inhibits glucose action even in the absence of glucose phosphorylation [11]. These data suggest that mannoheptulose acts on a molecule(s) distinct from glucokinase and inhibits insulin secretion. In other words, glucose is capable of stimulating insulin secretion in a condition without glucose metabolism, and mannoheptulose inhibits glucose-induced secretion in this condition by acting on other target molecules. Since mannoheptulose tastes sweet, it is an interesting possibility that it acts on the sweet taste receptor or related molecules. In this regard, *p*-nitrophenyl-D-glucopyranoside (PNP-Glu), which is known as an inhibitor of the sweet taste receptor in the tongue, has a strong inhibitory effect on the α anomer. Indeed, PNP-Glu also suppresses glucose-induced insulin secretion in pancreatic islets. It inhibits insulin secretion induced by the α-anomer in a concentration-dependent manner, but there is almost no inhibitory effect on the β-anomer. In addition, PNP-Glu does not affect insulin secretion induced by sulfonylurea or arginine [12]. It is an intriguing possibility that PNP-Glu acts on the sweet taste receptor or related molecules expressed in pancreatic β-cells and inhibits the glucose action. If so, this putative receptor functions as the glucoreceptor. It is known that alloxan is a glucose analog, and in rodents it selectively destroys pancreatic β-cells. Interestingly, the toxicity of alloxan is protected by co-administration of glucose. Also in this protective action, the two anomers have different effects. Thus, the α-anomer has a stronger protective effect compared to the β-anomer [13]. These findings strongly suggest that the site of action of alloxan is the glucoreceptor.

In 1988 Niki and colleagues investigated insulin secretion in animal models of diabetes. They isolated pancreatic islets from neonatal streptozotocin-administered rats (NSTZ rats) and Goto-Kakizaki rats (GK rats), two animal models of type 2 diabetes. When α- or β-anomer of glucose was administered, there was no difference in insulin secretion induced by the two anomers, which should have been observed in islets from wild type rats [14, 15]. Furthermore, Leclercq-Meyer and colleagues reported that the difference in insulin secretion induced by two the anomers disappeared in isolated pancreatic islets in BB rats, a model of type 1 diabetes [16]. The common phenotype for these diabetic model animals is not due to a change in the sensitivity to the β-anomer but rather to a decrease in the sensitivity to the α-anomer. Therefore, the portion of the glucoreceptor, recognizing that the α-anomer is a dominant stimulus, is more impaired than the one recognizing the β-anomer. In these animal models of diabetes, glucose metabolism in pancreatic islets does not differ from that of the wild type in glycolysis system, but it is still controversial as to the difference in the oxidation stage of glucose unrelated to anomeric discrimination.

At least a part of the dysfunction of pancreatic β-cells in diabetes is due to persistent exposure to hyperglycemia. However, sustained hyperglycemia does not down-regulate the activity of glycolytic enzymes. Furthermore, the function of glucokinase is regulated by expression, and hyperglycemia in fact increases the expression of glucokinase [17]. Hence, it is quite unlikely that hyperglycemia down-regulates glucose metabolism. In this regard, Niki and colleagues examined whether or not discrimination of glucose anomers was restored after normalization of plasma glucose levels in diabetic GK rats by administration of islet-activating protein (IAP). They found that pancreatic β-cells in IAP-treated GK rats recognize the α-anomer as a dominant stimulus as compared with the β-anomer [15]. Persistent hyperglycemia may have caused down-regulation of glucoreceptors and decreased the cognitive function of the anomers. These results indicate that the glucose-sensing mechanism is impaired in the diabetic condition and is restored by improving the blood glucose levels. Collectively, the function of the glucoreceptor is involved in the pathology of diabetes.

More than 40 years have already passed since the debates on "glucoreceptor hypothesis" and "metabolic hypothesis" have emerged. Since the discoveries of the K_{ATP} channel [18] and its molecular identification [19], "metabolic hypothesis" has become dominant. In addition, the molecular nature of the glucoreceptor was completely unknown in those days. Accordingly, it is not surprising that the "gluco-receptor hypothesis" became obscure.

However, numerous data suggesting the existence of "glucoreceptor" shown in the 1970s–90s are still valid. Unfortunately, the molecular nature of the "glucoreceptor" has not been clarified, yet many data suggest that it is a plasma membrane receptor molecule with a function similar to that of the sweet receptor in the taste bud [20, 21]. The research by Niki and colleagues may be the first act of glucoreceptor research, and elucidation of the molecular nature and function of the glucoreceptor is a big issue left for the second act.

References

1. Niki A, Niki H, Miwa I, Okuda J. Insulin secretion by anomers of d-glucose. Science. 1974;186:150–1.
2. Idahl LA, Sehlin J, Taljedal IB. Metabolic and insulin-releasing activities of D-glucose anomers. Nature. 1975;254:75–7.
3. Miwa I, Okuda J, Niki H, Niki A. Uptake of radioactive D-glucose anomers by pancreatic islets. J Biochem. 1975;78:1109–11.
4. Grodsky GM, Fanska R, West L, Manning M. Anomeric specificity of glucose-stimulated insulin release: evidence for a glucoreceptor? Science. 1974;186:536–8.
5. Bailey JM, Fishman PH, Pentchev PG. Studies on mutarotases. II. Investigations of possible rate-limiting anomerizations in glucose metabolism. J Biol Chem. 1968;234:4827–31.
6. Malaisse WJ, Malaisse-Lagae F, Sener A. Anomeric specificity of hexose metabolism in pancreatic islets. Physiol Rev. 1983;63:773–86.
7. Idahl LA, Rahemtulla F, Sehlin J, Taljedal IB. Further studies on the metabolism of D-glucose anomers in pancreatic islets. Diabetes. 1976;25:450–8.
8. Malaisse WJ, Sener A, Koser M, Herchuelz A. Identification of the alpha-stereospecific glucosensor in the pancreatic B-cell. FFBS Lett. 1976;65:131–4.
9. Malaisse WJ, Sener A, Koser M, Herchuelz A. Stimulus-secretion coupling of glucose-induced insulin release. Metabolism of alpha- and beta-D-glucose in isolated islets. J Biol Chem. 1976;251:5936–43.
10. Niki A, Niki H, Miwa I. Effect of anomers of D-mannose on insulin release from perfused rat pancreas. Endocrinology. 1979;105:1051–4.
11. Wolf BA, Florholmen J, Turk J, McDaniel ML. Studies of the Ca^{2+} requirements for glucose- and carbachol-induced augmentation of inositol trisphosphate and inositol tetrakisphosphate accumulation in digitonin-permeabilized islets. Evidence for a glucose recognition site in insulin secretion. J Biol Chem. 1988;263:3565–75.
12. Niki A, Niki H, Hashioka T. Receptors of paraneurons, with special reference to glucoreceptors. Arch Histol Cytol. 1989;52:suppl33–38.
13. Niki A, Niki H. Hexose anomers, insulin release, and diabetes mellitus. Biomed Res. 1980;1:189–206.
14. Niki A, Niki H, Niki I, Kunoh Y. Insulin release by glucose anomers in a rat model of non-insulin-dependent diabetes. Diabetologia. 1988;31:65–7.
15. Niki A, Niki H, Hashioka T, Niki I, Suzuki K, Goto Y. Insulin release by D-glucose anomers in a non-insulin-dependent diabetes rat model. Diabetes Res Clin Pract. 1989;7:87–92.
16. Leclercq-Meyer V, Marchand J, Malaisse WJ. Alteration of the insulin secretory response to D-glucose anomers in diabetic BB rats. Med Sci Res. 1987;15:1535–6.
17. Liang Y, Najafi H, Matschinsky FM. Glucose regulates glucokinase activity in cultured islets from rat pancreas. J Biol Chem. 1990;5:168663–6.

18. Ashcroft FM, Harrison DE, Ashcroft SJ. Glucose induces closure of single potassium channels in isolated rat pancreatic beta-cells. Nature. 1984;312(5993):446–8.
19. Inagaki N, Gonoi T, Clement JP 4th, Namba N, Gonzales G, Aguilar-Bryan L, Seino S, Bryan J. Reconstitution of IKATP: an inward rectifier subunit plus the sulfonylurea receptor. Science. 1995;270(5239):1166–70.
20. Nakagawa Y, Nagasawa M, Yamada S, Hara A, Mogami H, Nikolaev VO, Lohse MJ, Shigemura N, Ninomia Y, Kojima I. Sweet taste receptor expressed in pancreatic β cells activates the calcium and cyclic AMP signaling systems and stimulates insulin secretion. PLoS One. 2009;4:e5106.
21. Nakagawa Y, Ohtsu Y, Nagasawa M, Shibata H, Kojima I. Glucose promotes its own metabolism by acting on the cell-surface glucose-sensing receptor T1R3. Endocr J. 2014;61(2):119–31.

Chapter 3
KATP Channel-Independent Pathway and the Glucoreceptor

Toru Aizawa and Mitsuhisa Komatsu

Abstract The biphasic response is a unique characteristic of glucose-stimulated insulin secretion by islet beta cells. Namely, upon exposure to a high concentration of glucose, insulin secretion increases sharply for approximately 5 min before it gradually wanes at the end of the first phase. The second phase is characterized by sustained insulin release that lasts for the entire duration of high extracellular glucose concentration. Recently, new insights into the mechanisms underlying this phenomenon have been gained due to molecular identification of beta cell glucoreceptors. It has been previously postulated that elevation in cytosolic calcium leading to the first phase of insulin secretion is solely due to membrane depolarization caused by the closure of ATP-sensitive potassium (KATP) channels. However, the rapid increase in intracellular calcium is also caused by glucoreceptor-down signaling. Therefore, the first phase of glucose-induced secretion, i.e., fusion of the beta granule and the plasma membrane, likely arises from both glucose-receptor binding and closure of KATP channels. The molecular nature of the so-called KATP-independent glucose action appears to be diverse and still remains elusive. The releasable pool of beta granules is replenished by this glucose action, leading to the second phase of insulin secretion; signals for this phase also involve the direct recognition of glucose molecules by specific receptors. In this chapter, we review data pertaining to the functions of glucoreceptors in relation to KATP-independent glucose action in beta cells.

Keywords KATP channel · Biphasic response · Beta granule · Releasable pool · Cytosolic calcium

---

T. Aizawa (✉)
Diabetes Center, Aizawa Hospital, Honjo, Matsumoto, Japan
e-mail: taizawax@ai-hosp.or.jp

M. Komatsu
Department of Internal Medicine, Division of Diabetes, Endocrinology and Metabolism, Shinshu University School of Medicine, Matsumoto, Japan
e-mail: mitsuk@shinshu-u.ac.jp

I. Kojima (ed.), *Glucose-sensing Receptor in Pancreatic Beta-cells*,
https://doi.org/10.1007/978-981-13-0002-8_3

3.1 Introduction

In order for glucose-stimulated insulin secretion (GSIS) to occur, the glucose molecule has to be recognized by beta cells. Approximately 50 years ago, beta cell glucose recognition was reported to occur through two separate but additive mechanisms: the first through glucose metabolism and the other through beta cell surface glucoreceptors, such as in the tongue [1–4]. Following the identification of glucokinase (GK) as a rate limiter of glucose phosphorylation during glycolysis, GK was proposed as a beta cell glucose sensor [5]. However, as overall glucose metabolism by islet cells is dictated not at the level of glucose phosphorylation, but by a more distal step or steps, the GK glucose sensor paradigm was strongly opposed [6]. Nonetheless, for the past 30 years, the majority of beta cell investigators have been trying to identify the mechanism linking glucose metabolism and insulin release. The term metabolism-secretion coupling was coined in 1997 [7].

Nonetheless, survivors of the "glucoreceptor school" did existed. We identified ATP-sensitive K^+ channel (KATP)-independent stimulation of insulin secretion by glucose in 1992 [8–10], and we speculated that this glucose action is in part mediated by glucoreceptors [9] (Fig. 3.1). To be honest, this was in part due to intuition but mostly from a strong influence on us from late Professor Atsushi Niki, the world

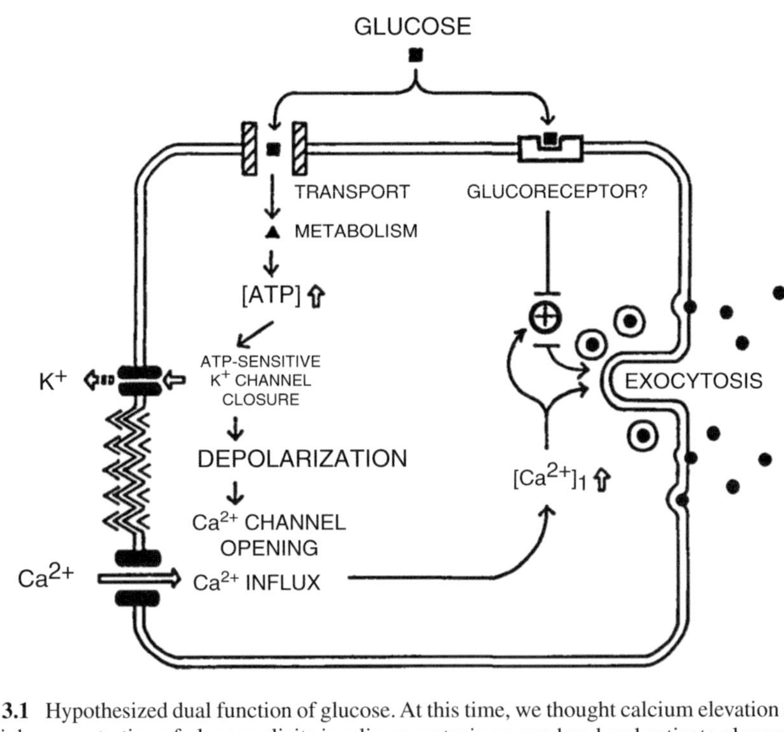

Fig. 3.1 Hypothesized dual function of glucose. At this time, we thought calcium elevation caused by high concentration of glucose elicits insulin exocytosis on one hand and activate glucoreceptor downstream signal(s). (Adapted from Ref. [8] with permission)

leader of the "glucoreceptor school." Recently, Dr. Kojima and colleagues have shown renewed interest on beta cell glucoreceptors [11–13]. In this chapter, we present a reappraisal of our perspective on beta cell glucoreceptor and its signaling pathway.

3.2 Glucose Stimulation of the Islet Beta Cell

A high concentration of glucose is the most important and robust physiological stimulus for insulin secretion by the beta cell, which elicits insulin exocytosis within minutes in vivo and even within a minute in vitro. In addition, a stimulatory concentration of glucose primes the beta cell, i.e., insulin release provoked by any stimuli such as nutrients, pharmacological agents, and forced membrane depolarization, which are all enhanced by a prior exposure to high glucose concentration. This glucose action is called time-dependent potentiation (TDP). The underlying mechanisms for the glucose triggering of insulin exocytosis and TDP are different. Namely, the former is largely dependent on glucose-induced KATP channel closure, and the latter is primarily independent of it. To easily understand the possible involvement of glucoreceptor in the KATP-independent glucose action, GSIS is first summarized below.

3.2.1 Insulin Secretion in Response to a High Concentration of Glucose

Upon elevation of its extracellular concentration, glucose enters the beta cell via the glucose transporter. Glucose is rapidly metabolized in the cell and the metabolic milieu of the beta cell changes. Among the variety of metabolic intermediates produced in the beta cell upon glucose stimulation, an increase in ATP and a decrease in ADP, and therefore increased ATP to ADP ratio is regarded as the major signal for the KATP channel closure (Fig. 3.2) [14]. Closure of the channel is followed by decreased K^+ outflow, which in turn causes membrane depolarization and the opening of L-type voltage-dependent Ca^{2+} channels. A rapid influx of calcium then ensues due to a large concentration gradient across the cell membrane; the extracellular free Ca^{2+} concentration is in the millimolar range, while intracellularly it is in the low nanomolar range. This Ca^{2+} influx raises the submembrane free Ca^{2+} concentration, which triggers a reaction cascade by activating the soluble N-ethylmaleimide-sensitive factor attachment protein receptor (SNARE) complex. For the fusion of beta granule to the plasma membrane. This cascade leads to a dramatic rise in the rate of fusion of the beta granules to the plasma membrane [15–17] (Fig. 3.3).

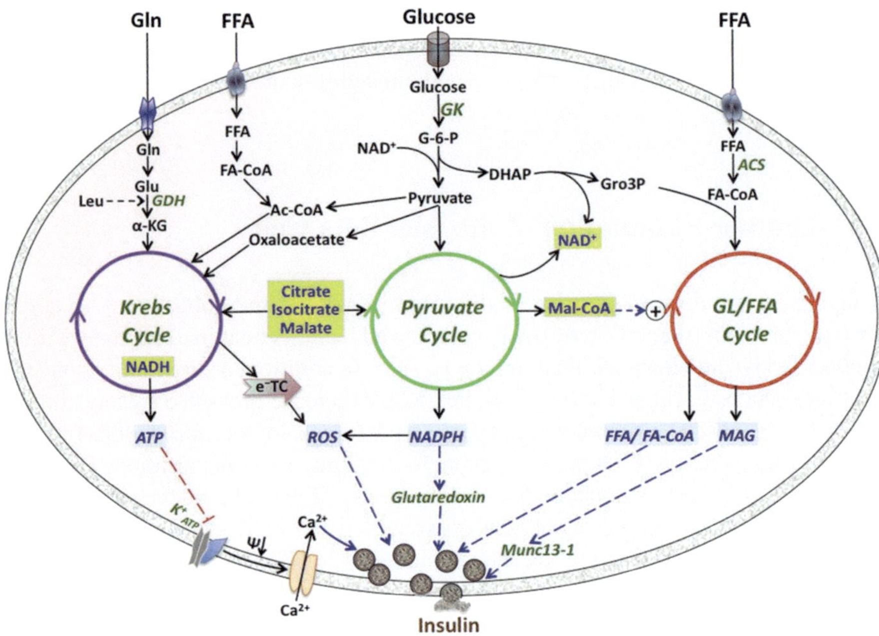

Fig. 3.2 Relationship between three fuel-driven metabolic cycles that generate metabolic coupling factors in the β cell. (Adapted from Ref. [14] with permission)

This concept was originally proposed by Dean and Mathews in 1968 [18] and has been confirmed in numerous studies performed thereafter [19]. This explanation of the action of glucose is straightforward and easy to follow. Furthermore, insulin secretion activated by glucose can be prevented or mimicked by inhibitors or activators of each step, respectively. For example, an activator of KATP channel (diazoxide), an inhibitor of voltage-dependent calcium channels (VDCC) (nifedipine), use of to the experimental buffer without added Ca^{2+} (note that this procedure lowers extracellular Ca^{2+} down to the low micromolar range, not lower than $[Ca^{2+}]i$), and inhibitors of glycolysis or mitochondrial metabolism all interfere with regular GSIS. On the other hand, an inhibitor of KATP channels (sulfonylurea), a VDCC activator (BAY-K 8644), and non-glucose fuel sources in the beta cell such as certain amino acids all stimulate insulin exocytosis. Even a simple increase of extracellular Ca^{2+} causes transient insulin release [20]. These observations are all compatible with the KATP dogma.

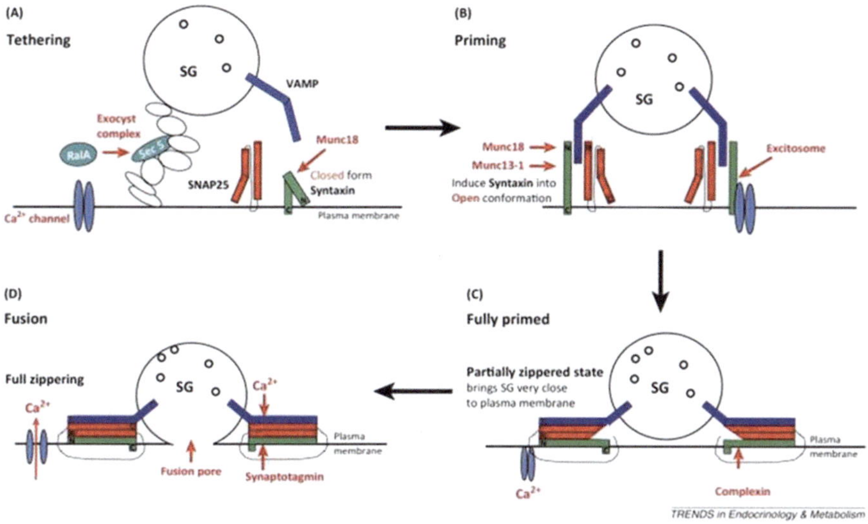

Fig. 3.3 Stages of exocytosis. (**a**) Secretory granule (SG) tethering to the plasma membrane (PM). RalA binds Sec5 to promote assembly of the eight-subunit exocyst complex, which tethers the SG loosely to the PM. Munc18 binds syntaxin in closed form, which is unable to form a complex with synaptosome-associated protein of 25 kDa(SNAP25) and vesicle-associated membrane protein (VAMP). (**b**) Priming. Munc18 is assisted by Munc13–1 in activating syntaxin, which adopts an open conformation that can form a soluble N-ethylmaleimide-sensitive factor attachment protein receptor (SNARE) complex with SNAP25 and VAMP. This SNARE complex is bound to Ca2+ channels to form the excitosome, which positions SGs at sites of maximum Ca2+ influx. (**c**) Fully primed. The SG is held close to the PM by the SNARE complex, which is only partially zippered (note the space at the C terminus) due to being clamped by complexin. (**d**) Fusion. Ca2+ influx from open Ca2+ channels (relieved from binding the SNARE complex) acts on the Ca2+ sensor synaptotagmin, which removes complexin from the SNARE complex thus allowing full zippering of the SNARE complex. Synaptotagmin also promotes lipid mixing between SGs and the PM, leading to fusion pore formation. (Adapted from Ref. [17] with permission)

3.2.2 Phasic Insulin Response: A Manifestation of KATP-Independent Glucose Action

Continued exposure of the islet beta cell to a stimulatory concentration of glucose elicits phasic insulin secretion [21]. The initial phase begins within a minute of stimulation and subsides at around 5–8 min, if the beta cell is continuously exposed to a consistent stimulatory concentration of glucose. The second phase involves a gradual increase in the rate of insulin exocytosis. This phase lasts for hours [21, 22] or essentially for as long as the beta cell is kept exposed to a high concentration of glucose. Importantly, the end of the first phase insulin release is not simply an emptying of the beta cell insulin reserve, as varying concentrations of stimulating glucose elicit a corresponding level of first phase response [21]. Obviously, the amount of insulin secreted in, for instance, 30-min incubation of the beta cell with a stimulatory concentration of glucose is mostly attributable to the second phase release. The second phase is relatively larger in rat and human islet beta cells than in mouse counterparts, which is considered in part due to difference in the metabolic fate of glucose in rat and mouse beta cells [23]. Terminologies such as "augmentation," "amplification," and "the second phase" have been interchangeably used by investigators in this field [19, 24].

The temporal profile of KATP channel closure, membrane depolarization, and elevation of cytosolic Ca^{2+} do not match to the phasic insulin response as described above [25]. Glucose can induce second phase insulin secretion in the presence of diazoxide, a KATP channel opener, if cytosolic Ca^{2+} is elevated by a depolarizing concentration of K^+ [8] or a VDCC opener, Bay-K 8644 [26]. Glucose induces Ca^{2+} oscillation in the cell, but simple depolarization by high concentration of K^+ does not. Therefore, the Ca^{2+} oscillation was proposed as a putative drive for the second phase insulin release. However, the glucose-induced second phase insulin release is present under full opening of KATP channels [8], which denies the legitimacy of this hypothesis. Selective inhibition of the first phase does not affect the amplitude or temporal profile of the second phase response [25]. Therefore, this so-called second phase insulin secretion in response to glucose is attributable to KATP-independent mechanism or mechanisms.

3.2.3 Time-Dependent Potentiation: An Alternative Manifestation of KATP-Independent Glucose Action

When islet beta cells are preexposed to a stimulatory concentration of glucose, they secrete a greater amount of insulin than those not preexposed, in response to stimuli applied later. This phenomenon is called time-dependent potentiation (TDP), priming, or glucose memory [26–28]. It has been well established that not only glucose, but other nutrients, which are metabolized in beta cells, elicit TDP. Certain amino acids and free fatty acids are representative of such nutrients. On the other hand, ionic events, such as depolarization or elevation of cytosolic $[Ca^{2+}]i$ or activation of cAMP-dependent processes by incretin hormones, do not cause TDP if nutrients were not present. Activation of protein kinase C may be an exception; pretreatment of the beta cell with an activator of protein kinase C, such as phorbol ester, causes TDP even in the absence of a stimulatory concentration of glucose [29].

Independence of TDP from the KATP channel was first phenomenologically shown by Grill et al. [30]. Here, they elegantly demonstrated that glucose-induced TDP occurs even if glucose priming was performed in the presence of a high concentration of diazoxide; diazoxide was not considered as an opener of the KATP channel, however. We established that TDP is totally independent of KATP channel closure and therefore independent of an elevation of cytosolic Ca^{2+} [26, 27]. Moreover, TDP occurs seemingly normal in the buffer without added calcium and containing 1 mM EGTA [28].

3.2.4 "Fusion" Rather than Triggering and "Replenishment" in Place of Augmentation/Amplification: "The Inside Story"

Insulin secretion by the islet beta cell has been quantified primarily by insulin immunoassay, by which amount of insulin molecule coming out of the cell was measured. The term "triggering and amplification/augmentation" [19, 24] has stemmed from such measurements, which we would call "the outside story." In such studies, the process of glucose-stimulated insulin secretion is "interpreted" from the cell exterior to speculate what is occurring intracellularly. More recently, the intracellular dynamics of beta cells has increasing been elucidated by direct visualization of beta granule trafficking [17]. Previously, we proposed a scheme in which the traditional, indirect "outside" view of phasic insulin secretion was replaced with a more direct inside view [16]. Here, we push further and present a more comprehensive scheme encompassing the glucoreceptor and its downstream signaling (Fig. 3.4).

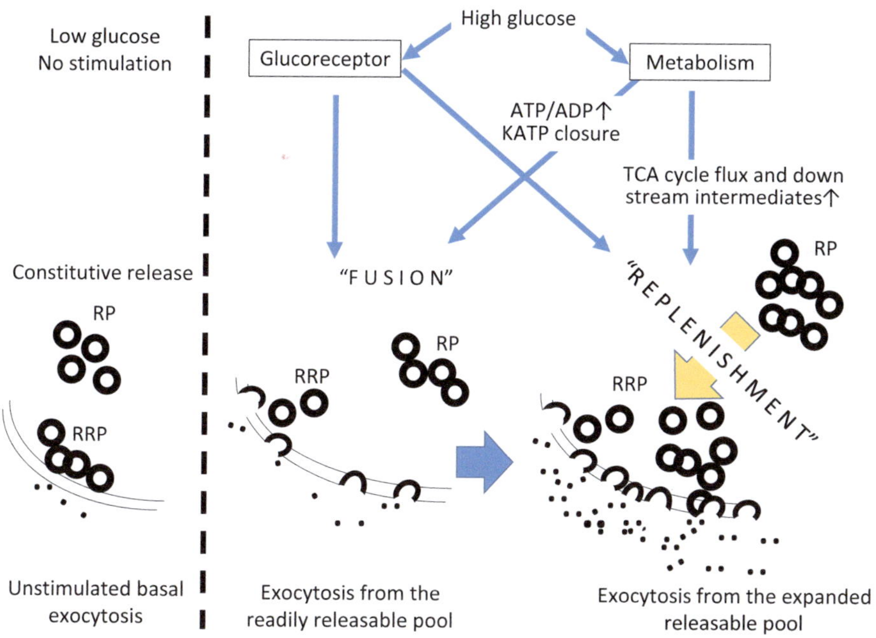

Fig. 3.4 Phasic insulin secretion upon stimulation with high concentration of glucose. Putative role of glucoreceptor-down signaling was illustrated in addition to KATP-dependent and KATP-independent glucose actions. See Text for the detail

The beta granule is not homogeneous. Only a small proportion of insulin in the cell (15% maximum) is packaged in the releasable pool of the granule [27], that is, the readily releasable (RRP) and remote pool (RP) combined. Namely, the majority of insulin resides in non-releasable pool(s) [50]. The so-called first phase insulin secretion, or triggered insulin, consists of insulin from the readily releasable pool, through rapid membrane fusion of the beta granule and the plasma membrane (PM). Concurrently, replenishment/refilling of the readily releasable pool occurs. This replenishment may be in part due to physical movement of the beta granule into the pools closer to the PM but can primarily be attributed to chemical modification of the granule surface SNARE complex rendering them to be readily releasable. In other words, there may be a mature and immature beta granule with the former being the first (immediate) comers and the latter late (slow) comers [14, 17].

Second phase insulin secretion, amplification, augmentation, and TDP could all be regarded as a manifestation of replenishment. Specifically, "second phase of GSIS," "amplification," "augmentation," and "time-dependent potentiation" are different "outside" names for the same "inside" event, which is the replenishment of the RRP from the RP. This process is entirely governed by the KATP-independent glucose action.

3.3 The Glucoreceptor Hypothesis and the KATP Channel-Independent Glucose Action

The involvement of glucoreceptors for KATP channel-independent glucose action was speculated by our group in 1992 [9]. Subsequently, we have evaluated this hypothesis by several different modalities, including pharmacological inhibition of the sweet taste inhibitor, the metabolic inhibition of beta cells, and the stringent Ca^{2+} removal from the media.

3.3.1 Impact of Pharmacological Inhibition of Glucoreceptors on the KATP Channel-Independent Glucose Action

Sweet substances including glucose are recognized by the cell surface glucoreceptor of sweet sensing cells of the taste bud. As para-nitrophenyl-alpha-D-glucopyranoside is a blocker of glucoreceptors in the tongue, its effect on the KATP channel-independent glucose action was tested [31]. A high concentration of glucose robustly enhanced insulin release, evoked by a depolarizing concentration of K^+ in the presence of diazoxide and a KATP channel opener, which is the core of KATP-independent glucose action.

Pretreatment of the rat islet cells for 30 min with 5–10 mM p-nitrophenyl-alpha-D-glucopyranoside, but not its beta isomer, significantly suppressed (by 20–60%) such glucose action. Conversely, the sweet taste inhibitor did not suppress insulin release induced by a depolarizing concentration of K^+ in the absence of a stimulatory concentration of glucose. The sweet taste inhibitor did not suppress glucose metabolism in the beta cell [31]. This was evidence indicating that KATP channel-independent glucose action is in part mediated by glucoreceptors.

3.3.2 Resistance to Metabolic Inhibition of KATP Channel-Independent Glucose Action

It is well established that glucose metabolism in the beta cell is attenuated by prolonged fasting of animals or by the lowering of incubation temperature. We utilized this unique feature of beta cells to prove the relative independence of the KATP channel-independent glucose action on glucose metabolism [31, 32]. First, as in previous experiments, the KATP-independent glucose action was confirmed in the release experiment where glucose was added on top of depolarizing concentration of K^+ in the presence of diazoxide, an activator of the KATP channel. Islet beta cells derived from rats fasted for 4 days prior to the experiment showed almost no insulin secretion in response to 16.7 mmol/l glucose. Alternatively, the KATP channel-independent glucose-mediated stimulation of insulin secretion was only 50% lower compared to that seen in normally fed rats' beta cells. Under a low temperature of 22 °C, insulin release induced by 16.7 mmol/l glucose alone was negligible. In contrast, KATP-independent glucose action was clearly detectable at this low temperature. Glucose metabolism, glycolysis, and glucose oxidation in the islet cells were suppressed by 20% and 50%, respectively, after 4 days fasting. At a low temperature, both glycolysis and glucose oxidation were lowered by 60% [31, 32].

Such KATP channel-independent glucose action can be demonstrated even in stringent calcium-free conditions [33]. Namely, even after incubation in the buffer without added calcium and 1 mM EGTA present, glucose robustly stimulated insulin secretion if TPA, a classic protein kinase C activator, was present [33]. In addition, the rapid oscillation of insulin release from single islets upon glucose stimulation was present even under these stringent Ca^{2+} free conditions [34]. It has been reported that glycolysis and glucose oxidation were negligible in stringently Ca^{2+}-free media [35]. Rapid oscillatory, pulsatile insulin release in response to high glucose was also retained despite KATP channels were fully open [36].

Therefore, KATP-independent glucose stimulation of insulin release is resistant to the inhibition of metabolism, which again indicates the role of glucoreceptor in this branch of glucose action.

3.3.3 KATP-Independent Glucose Action in Patients or Experimental Animals with Genetic Loss of a Functional KATP Channel

The most convincing evidence so far to support the proposed functioning of the glucoreceptor in human beta cells is the following observation in patients with genetic abnormality resulting in the absence of KATP channel: persistent hyperinsulinemic hypoglycemia of infancy (PHHI) [37]. An abrupt rise in plasma glucose concentration from 75 to 350 mg/dl (4.2–19.4 mM) through intravenous bolus glucose injection to the children with this mutation elicited a near normal insulin secretion. The absence of a functional KATP channel was proven both by genetic analysis of ex vivo pancreatic tissue and by the absence of insulin secretion upon injection of a pharmacological dose of sulfonylurea, a KATP inhibitor insulin secretagogue used for patients with diabetes. In vivo, islet beta cells are kept primed with an ambient level of incretins, amino acids, fatty acids, and parasympathetic input. Instantaneous elevation of glucose to the above-described level may be able to cause a near-normal insulin secretion, despite the absence of KATP channels. Additionally, loss of insulin response to IV bolus glucose is an early abnormality of the islet beta cell [38]. In unison, the above findings are of prime importance, strongly suggesting the KATP-independent, possibly glucoreceptor-mediated mechanism leading to insulin exocytosis, is defective in human diabetes. Subtotal pancreatectomy in patients with genetic loss of KATP channels resulted in normal glucose metabolism or mild diabetes that lasted for years [39–41]. Such clinical data implies that nutrient-regulated insulin secretion occurs in vivo in the absence of the beta cell KATP channels. Pancreatic islets from PHHI patients responded well to glucose in vitro [42].

The data obtained from the KATP channel knockout (KO) mouse are variable and highly discordant to each other. Namely, in islet beta cells from this KATP channel KO mouse, glucose-induced insulin release was absent, impaired, or retained. In one strain, KO of the KATP channel obliterated GSIS in vivo [43]. On the other hand, GSIS by the beta cell was clearly retained [44] or close to normal [45] in the KATP-KO mouse of other strains. Recovery of glucose responsiveness of the islets from KATP channel KO mouse by overnight culture was demonstrated [45]. The inconsistency of phenotype of the KATP channel KO mouse might be due to genetic background of the parent mouse. There are two subunits for the KATP channel: one an inward rectifying potassium channel 6.2 (Kir 6.2) and the other a sulfonylurea receptor 1 (SUR1) [46]. Total obliteration of GSIS by the islet beta cell has been reported in mice with *Kir6.2* KO [43]. On the other hand, retention of GSIS has been found in *SUR1* KO mouse [44, 45, 47] and in the beta cells from PHHI patients with *SUR1* mutation [42]. Considering that the regulatory/accessory proteins for Kir6.2 and SUR1 are totally different, it might be possible that the phenotypic difference of the two groups of KATP-KO mice is attributable to the distinct set of proteins wiped out by the *Kir6.2* and *SUR1* deletion (Prof. Kakei, personal communication).

3.4 Summary

For over 80 years, the mechanism of glucose-mediated stimulation of insulin secretion has been studied by many researchers using divergent techniques. However, we do not yet fully understand it. Specifically, the mediator(s) of KATP-independent glucose action, namely, factor(s) mediating glucose replenishment of the RRP, are still ambiguous [48, 49]. Glucoreceptors may well be an important player in glucose stimulation of beta cells. Particularly, it may function as a bridge among ionic, non-ionic, and metabolic signaling pathways and is needed for the maximum manifestation of glucose stimulation of the beta cell.

References

1. Matschinsky FM, Landgraf R, Ellerman J, Kotler-Brajtburg J. Glucoreceptor mechanisms in islets of Langerhans. Diabetes. 1972;21(Suppl 2):555–69.
2. Matschinsky FM, Pagliara AS, Hover BA, Haymond MW, Stillings SN. Differential effects of alpha- and beta-D- glucose on insulin and glucagon secretion from the isolated perfused rat pancreas. Diabetes. 1975;24:369–72.
3. Ashcroft SJH, Bassett JM, Randle PJ. Insulin secretion mechanisms and glucose metabolism in isolated islets. Diabetes. 1972;21(Suppl 2):538–45.
4. Niki A, Niki H, Miwa I, Okuda J. Insulin secretion by anomers of D-glucose. Science. 1974;186:150–1.
5. Bedoya FJ, Wilson JM, Ghosh AK, Finegold D, Matschinsky FM. The glucokinase glucose sensor in human pancreatic islet tissue. Diabetes. 1986;35:61–7.
6. Malaisse WJ, Sener A. Glucokinase is not the pancreatic B-cell glucoreceptor. Diabetologia. 1985;28:520–7.
7. Maechler P, Kennedy ED, Pozzan T, Wollheim CB. Mitochondrial activation directly triggers the exocytosis of insulin in permeabilized pancreatic beta-cells. EMBO J. 1997;16:3833–41.
8. Sato Y, Aizawa T, Komatsu M, Okada N, Yamada T. Dual functional role of membrane depolarization/Ca^{2+} influx in rat pancreatic B-cell. Diabetes. 1992;41:438–43.
9. Aizawa T, Sato Y, Komatsu M, Hashizume K. ATP-sensitive K^+ channel-independent, insulinotropic action of glucose in the B-cells. Endocr Regul. 1992;26:159–62.
10. Gembal M, Gilon P, Henquin JC. Evidence that glucose can control insulin release independently from its action on ATP-sensitive K^+ channels in mouse B cells. J Clin Invest. 1992;89:1288–95.
11. Kojima I, Nakagawa Y, Hamano K, Medina J, Li L, Nagasawa M. Glucose-sensing receptor T1R3: a new signaling receptor activated by glucose in pancreatic β-cells. Biol Pharm Bull. 2015;38:674–9.
12. Nakagawa Y, Ohtsu Y, Nagasawa M, Shibata H, Kojima I. Glucose promotes its own metabolism by acting on the cell-surface glucose-sensing receptor T1R3. Endocr J. 2014;61:119–31.
13. Hamano K, Nakagawa Y, Ohtsu Y, Li L, Medina J, Tanaka Y, Masuda K, Komatsu M, Kojima I. Lactisole inhibits the glucose-sensing receptor T1R3 expressed in mouse pancreatic β-cells. J Endocrinol. 2015;226:57–66.
14. Prentki M, Matschinsky FM, Madiraju SR. Metabolic signaling in fuel-induced insulin secretion. Cell Metab. 2013;18:162–85.
15. Nagamatsu S, Ohara-Imaizumi M. Imaging exocytosis of single insulin secretory granules with TIRF microscopy. Methods Mol Biol. 2008;440:259–68.

16. Aizawa T, Komatsu M. Rab27a: a new face in beta cell metabolism-secretion coupling. J Clin Invest. 2005;115:227–30.
17. Gaisano HY. Here come the newcomer granules, better late than never. Trends Endocrinol Metab. 2014;25(8):381.
18. Dean PM, Matthews EK. Electrical activity in pancreatic islet cells. Nature. 1968;219:389–90.
19. Straub SG, Sharp GW. Glucose-stimulated signaling pathways in biphasic insulin secretion. Diabetes Metab Res Rev. 2002;18:451–63.
20. Devis G, Somers G, Malaisse WJ. Dynamics of calcium-induced insulin release. Diabetologia. 1977;13:531–6.
21. Grodsky GM. Threshold distribution hypothesis for packet storage of insulin. II. Effect of calcium. Diabetes. 1972;21(Suppl 2):584–93.
22. Bolaffi JL, Heldt A, Lewis LD, Grodsky GM. The third phase of in vitro insulin secretion: evidence for glucose insensitivity. Diabetes. 1986;35:370–21.
23. MacDonald MJ. Differences between mouse and rat pancreatic islets: succinate responsiveness, malic enzyme, and anaplerosis. Am J Physiol Endocrinol Metab. 2002;283:E302–10.
24. Henquin JC. The dual control of insulin secretion by glucose involves triggering and amplifying pathways in β-cells. Diabetes Res Clin Pract. 2011;93(Suppl 1):S27–31.
25. Kaminski MT, Lenzen S, Baltrusch S. Real-time analysis of intracellular glucose and calcium in pancreatic beta cells by fluorescence microscopy. Biochim Biophys Acta. 2012;1823:1697–707.
26. Taguchi N, Aizawa T, Sato Y, Ishihara F, Hashizume K. Mechanism of glucose-induced biphasic insulin release: physiological role of adenosine triphosphate-sensitive K^+ channel-independent glucose action. Endocrinology. 1995;136:3942–8.
27. Yamada S, Komatsu M, Aizawa T, Sato Y, Yajima H, Yada T, Hashiguchi S, Yamauchi K, Hashizume K. Time-dependent potentiation of the beta cell, a Ca^{2+}-independent phenomenon. J Endocrinol. 2002;172:345–54.
28. Yamada S, Komatsu M, Sato Y, Yamauchi K, Kojima I, Aizawa T, Hashizume K. Time-dependent stimulation of insulin exocytosis by 3′,5′-cyclic adenosine monophosphate in the rat islet beta-cell. Endocrinology. 2002;143:4203–9.
29. Niki I, Tamagawa T, Niki H, Niki A, Koide T, Sakamoto N. Possible involvement of diacylglycerol-activated, Ca^{2+}-dependent protein kinase in glucose memory of the rat pancreatic B-cell. Acta Endocrinol. 1988;118:204–8.
30. Grill V, Adamson U, Rundfeldt M, Andersson S, Cerasi E. Glucose memory of pancreatic B and A2 cells evidence for common time-dependent actions of glucose on insulin and glucagon secretion in the perfused rat pancreas. J Clin Invest. 1979;64:700–7.
31. Aizawa T, Sato Y, Ishihara F, Komatsu M, Taguchi N, Hashizume K, Yamada T. ATP-sensitive K^+ channel-independent glucose action in rat pancreatic s-cell. Am J Phys. 1994;266:C622–7.
32. Ishihara F, Aizawa T, Taguchi N, Sato Y, Hashizume K. Differential metabolic requirement for initiation and augmentation of insulin release by glucose: a study with rat pancreatic islets. J Endocrinol. 1994;143:497–503.
33. Komatsu M, Schermerhorn T, Aizawa T, Sharp GWG. Glucose stimulation of insulin release in the absence of extracellular Ca^{2+} and in the absence of any rise in intracellular Ca^{2+} in rat pancreatic islets. Proc Natl Acad Sci U S A. 1995;92:10728–32.
34. Aizawa T, Kaneko T, Yajima H, Yamada S, Sato Y, Kanda Y, Kanda S, Noda M, Kadowaki T, Nagai M, Yamauchi K, Komatsu M, Hashizume K. Rapid oscillation of insulin release by the rat pancreatic islets under stringent Ca^{2+}-free conditions. J Endocrinol. 2000;166:545–51.
35. Denton RM. Regulation of mitochondrial dehydrogenases by calcium ions. Biochim Biophys Acta. 2009;1787:1309–16.
36. Westerlund J, Ortsäter H, Palm F, Sundsten T, Bergsten P. Glucose-regulated pulsatile insulin release from mouse islets via the K(ATP) channel-independent pathway. Eur J Endocrinol. 2001;144:667–75.
37. Grimberg A, Ferry RJ, Jr KA, Koo-McCoy S, Polonsky K, Glaser B, Permutt MA, Aguilar-Bryan L, Stafford D, Thornton PS, Baker L, Stanley CA. Dysregulation of insulin secretion

in children with congenital hyperinsulinism due to sulfonylurea receptor mutations. Diabetes. 2001;50:322–8.
38. Ward WK, Bolgiano DC, McKnight B, Halter JB, Porte D Jr. Diminished B cell secretory capacity in patients with noninsulin-dependent diabetes mellitus. J Clin Invest. 1984;74:1318–28.
39. Straub SG, Cosgrove KE, Ammälä C, Shepherd RM, O'Brien RE, Barnes PD, Kuchinski N, Chapman JC, Schaeppi M, Glaser B, Lindley KJ, Sharp GW, Aynsley-Green A, Dunne MJ. Hyperinsulinism of infancy: the regulated release of insulin by KATP channel-independent pathways. Diabetes. 2001;50:329–39.
40. Soliman AT, Alsalmi I, Darwish A, Asfour MG. Growth and endocrine function after near total pancreatectomy for hyperinsulinaemic hypoglycaemia. Arch Dis Child. 1996;74:379–85.
41. Ismail D, Werther G. Persistent hyperinsulinaemic hypoglycaemia of infancy: 15 years' experience at the Royal Children's Hospital (RCH), Melbourne. J Pediatr Endocrinol Metab. 2005;18:1103–9.
42. Tyrrell VJ, Ambler GR, Yeow WH, Cowell CT, Silink M. Ten years' experience of persistent hyperinsulinaemic hypoglycaemia of infancy. J Paediatr Child Health. 2001;37:483–8.
43. Miki T, Nagashima H, Tashiro F, Kotake K, Yoshitomi H, Tamamoto A, Gonoi T, Iwanaga T, Miyazaki JI, Seino S. Defective insulin secretion and enhanced insulin action in K_{ATP} channel-deficient mice. Proc Natl Acad Sci U S A. 1998;95:10402–6.
44. Seghers V, Nakazaki M, DeMayo F, Aguilar-Bryan L, Bryan J. Sur1 knockout mice. A model for K(ATP) channel-independent regulation of insulin secretion. J Biol Chem. 2000;31(275):9270–7.
45. Nenquin M, Szollosi A, Aguilar-Bryan L, Bryan J, Henquin JC. Both triggering and amplifying pathways contribute to fuel-induced insulin secretion in the absence of sulfonylurea receptor-1 in pancreatic beta-cells. J Biol Chem. 2004;279:32316–24.
46. Inagaki N, Gonoi T, Clement JP 4th, Namba N, Inazawa J, Gonzalez G, Aguilar-Bryan L, Seino S, Bryan J. Reconstitution of IKATP: an inward rectifier subunit plus the sulfonylurea receptor. Science. 1995;270:1166–70.
47. Shiota C, Larsson O, Shelton KD, Shiota M, Efanov AM, Hoy M, Lindner J, Kooptiwut S, Juntti-Berggren L, Gromada J, Berggren PO, Magnuson MA. Sulfonylurea receptor type 1 knock-out mice have intact feeding-stimulated insulin secretion despite marked impairment in their response to glucose. J Biol Chem. 2002;277:37176–83.
48. Ferdaoussi M, MacDonald PE. Toward connecting metabolism to the exocytotic site. Trends Cell Biol. 2017;27:163–71.
49. Kalwat MA, Cobb MH. Mechanisms of the amplifying pathway of insulin secretion in the β cell. Pharmacol Ther. 2017;179:17–30. [Epub ahead of print]
50. Komatsu M, Aizawa T, Yokokawa N, Sato Y, Okada N, Takasu N, Yamada T. Mastoparan-induced hormone release from rat pancreatic islets. Endocrinology. 1992;130:221–8.

Chapter 4
Signaling System Activated by the Glucose-Sensing Receptor

Johan Medina and Yuko Nakagawa

Abstract Glucose is a primary stimulator of insulin secretion in pancreatic β-cells. It has long been thought that glucose augments insulin secretion solely by a mechanism dependent on glucose metabolism. Consequently, it takes a certain period of time for glucose to initiate cellular responses. With regard to the membrane potential, for example, it takes at least half a minute to observe glucose-induced depolarization of the plasma membrane. This lag period is thought to be a time required for glucose metabolism. To address the possibility that glucose activates a cell-surface receptor, we developed sensitive methods to monitor changes in cytoplasmic free calcium ($[Ca^{2+}]_c$), cyclic AMP ($[cAMP]_c$), and activation of protein kinase C (PKC). Using sensitive methods, we investigated whether or not glucose induces immediate signals in β-cells. Indeed, glucose evoked immediate changes in $[Ca^{2+}]_c$, $[cAMP]_c$ and PKC activity. Importantly, these rapid signals were independent of glucose metabolism and were reproduced by addition of nonmetabolizable glucose analogs. Since these signals were inhibited by inhibition of Gq or Gs, it is quite likely that glucose activates a cell-surface receptor and generates immediate intracellular signals in pancreatic β-cells.

Keywords Glucose · Insulin secretion · Glucose-sensing receptor · T1R3 · Calcium-sensing receptor · Calcium · Cyclic AMP · Protein kinase C

4.1 Introduction

Insulin is a major regulator of glucose metabolism and is secreted from pancreatic β-cells [1]. Secretion of insulin is modulated by various regulators including nutrients, hormones, and neurotransmitters [1, 2]. Among them, glucose is the most important stimulator of insulin secretion. The action of glucose is unique in that it is able to stimulate insulin secretion by itself, and because of this, glucose is thought

J. Medina · Y. Nakagawa (✉)
Institute for Molecular and Cellular Regulation, Gunma University, Maebashi, Japan
e-mail: j.medina@uq.edu.au; nakayuko@gunma-u.ac.jp

I. Kojima (ed.), *Glucose-sensing Receptor in Pancreatic Beta-cells*,
https://doi.org/10.1007/978-981-13-0002-8_4

to be a primary stimulator of insulin secretion [1]. Many researchers have been studying the mechanism by which glucose stimulates insulin secretion. Dean and Mathew [3] found in their electrophysiological study that glucose induces action potentials in β-cells, which are dependent on extracellular Ca^{2+}. Subsequent studies have revealed that permeability of potassium is reduced by an elevation of ambient glucose concentration [4]. In 1984, Aschroft and colleagues [5] found the ATP-sensitive potassium channel (K_{ATP} channel) in pancreatic β-cells. Their results indicate that glucose is metabolized in β-cells, and a resultant increase in ATP or ATP/ADP ratio inhibits the K_{ATP} channel, which leads to depolarization of the plasma membrane. When the membrane potential exceeds the threshold, the voltage-dependent Ca^{2+} channel is opened and Ca^{2+} enters the cell [6]. These changes in ion fluxes are caused by changes in ATP, a product of glucose metabolism. Accordingly, these results led to the idea that glucose exerts its action by a mechanism dependent on its metabolism [7, 8].

4.2 Action of Glucose on Cellular Ca^{2+} Metabolism

As mentioned above, glucose induces changes in electrical activity in pancreatic β-cells [3]. Specifically, glucose causes depolarization of the plasma membrane. It should be mentioned that glucose-induced depolarization takes place after a certain lag time and it takes 1 min or sometimes longer. Following this, there are a series of action potentials, which is dependent on extracellular calcium. It has been generally thought that the lag period corresponds to the time required for generation of ATP. In other words, it takes a certain amount of time for glucose to be metabolized through the glycolytic pathway and in the mitochondria. An increase in the ATP/ADP ratio causes closure of the K_{ATP} channel and depolarizes the plasma membrane [5].

When changes in cytoplasmic-free calcium concentration ($[Ca^{2+}]_c$) are monitored by using a fluorescent Ca^{2+} indicator, for example, Fura-2, elevation of ambient glucose induces rather unexpected responses. Thus, a high concentration of glucose first causes reduction of $[Ca^{2+}]_c$ [9, 10]. This reduction of $[Ca^{2+}]_c$ lasts 50–60 s or even more, which is followed by oscillatory elevations of $[Ca^{2+}]_c$. Typically, oscillation of $[Ca^{2+}]_c$ lasts as long as the time it takes glucose to stimulate β-cells. Generally speaking, calcium mobilizing agonists induce an immediate elevation of $[Ca^{2+}]_c$ and, in β-cells, a muscarinic agonist carbachol, for example, evokes an immediate elevation of $[Ca^{2+}]_c$ [11]. Accordingly, the reduction of $[Ca^{2+}]_c$ induced by a high concentration of glucose is rather unexpected. Many researchers have been studying the mechanism by which glucose first reduces $[Ca^{2+}]_c$ in β-cells. They have found that the initial reduction of $[Ca^{2+}]_c$ is due to uptake of Ca^{2+} into intracellular organelle and that the endoplasmic reticulum (ER) may be the major site to which Ca^{2+} is taken up [12, 13]. Thus, glucose activates the Ca^{2+} pump in ER and thereby extrudes cytoplasmic Ca^{2+} into ER. This is the initial action of glucose on cellular Ca^{2+}. Although many studies have focused on the initial action of glucose on intracellular Ca^{2+}, it is still unclear how glucose activates the Ca^{2+} pump in ER

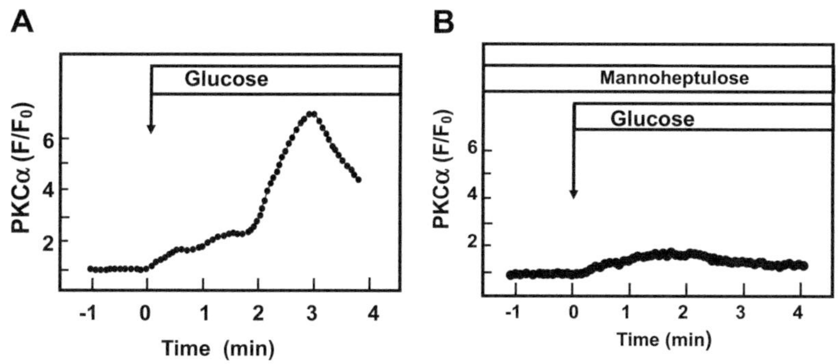

Fig. 4.1 Effect of glucose on translocation of PKC. (**a**) PKCα-expressing MIN6 cell was stimulated by 16.7 mM glucose and translocation of PKCα was monitored by TIRF microscopy. (**b**) Effect of glucose was examined as in **a** in the presence of 10 mM mannoheptulose

and reduces $[Ca^{2+}]_c$. Also, the physiological significance of the reduction of $[Ca^{2+}]_c$ is not totally clear. Furthermore, it is even uncertain whether or not the initial reduction of $[Ca^{2+}]_c$ is dependent on glucose metabolism. Taken together, the rapid actions of glucose in β-cells still remain elusive.

4.3 Rapid Signals Induced by Glucose

We were interested in the action of glucose in pancreatic β-cells and specifically interested in the initial action of glucose. Our research involved monitoring various changes in intracellular signals in a living β-cell evoked by high concentrations of glucose. To this end, we first developed a method to monitor the activation process of protein kinase C (PKC) by using total internal reflection fluorescence (TIRF) microscopy [14]. This is a very sensitive method to monitor translocation of PKC-α from cytosol to the plasma membrane. Conventional PKC including PKC-α is located in the cytosol in unstimulated conditions. When activated by various Ca^{2+}-mobilizing agonists, conventional PKC translocates to the plasma membrane. PKC is then released from the plasma membrane to the cytosol when the agonist signal is terminated. Accordingly, the activation process of PKC can be observed by monitoring the changes in the amount of PKC in the plasma membrane. As shown in Fig 4.1a, when a β-cell is stimulated by a high concentration of glucose, there is a small and gradual elevation of the amount of PKCα. This response is rapid and is detected within 10 s of the stimulation by glucose. A few min later, there is a large elevation of the amount of PKC in the plasma membrane. Considering the time frame of changes in $[Ca^{2+}]_c$ in glucose-stimulated β-cells, the second large elevation of PKC may correspond to the second phase of $[Ca^{2+}]_c$ response. A critical question is whether or not these changes in PKC are dependent on glucose metabolism. Glucose metabolism can be blocked by adding mannoheptulose, an inhibitor of

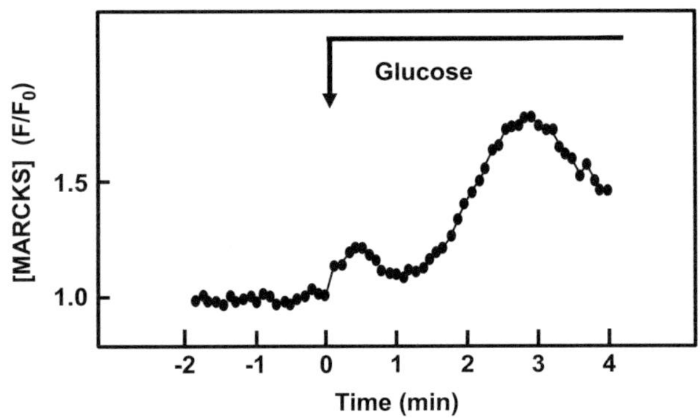

Fig. 4.2 Effect of glucose on phosphorylation of MARCKS. GFP-MARCKS-expressing MIN6 cell was stimulated by 16.7 mM glucose and changes in the amount of GFP-MARCKS in cytosol was monitored

glucokinase [15]. In the presence of mannoheptulose, the rapid gradual elevation of PKC is observed, whereas the second oscillatory elevation of PKC is completely blocked (Fig 4.1b). This result indicates that the first response of PKC is independent of glucose metabolism, while the second oscillatory elevation of PKC is dependent on glucose metabolism. For the first time metabolism-independent signals evoked by glucose can be observed. To confirm this, we administered nonmetabolizable glucose analog 3-*O*-methylglucose (3OMG). If the signal is produced by a mechanism independent of glucose metabolism, it would be expected that 3OMG is able to reproduce the signal. Indeed, 3OMG reproduces the first gradual elevation of PKC in the plasma membrane. In contrast, the second oscillatory elevation of PKC is not observed. These results suggest that the first response of PKC induced by glucose is independent of glucose metabolism. Since the first response occurs rapidly, the first response of PKC may be due to the activation of cell-surface receptor for glucose. We tentatively named this putative receptor, glucose-sensing receptor (GSR) [16].

To confirm the rapid activation of PKC induced by glucose, we monitored phosphorylation of myristoylated alanine-rich C-kinase substrate (MARCKS) [17]. In an unstimulated condition, MARCKS is bound to the plasma membrane. When phosphorylated by PKC, MARCKS is released to the cytosol [18]. Therefore, the phosphorylation state of MARCKS can be observed by monitoring the changes in the amount of MARCKS in the cytosol. As shown in Fig 4.2, administration of glucose induces biphasic elevation of MARCKS in the cytosol. Glucose induces a rapid transient elevation of MARCKS, which is followed by the second larger elevation of MARCKS in the cytosol. Interestingly, inhibition of glucose metabolism by administration of mannoheptulose abolishes the second response of MARCKS, whereas the first response is not affected by the inhibition of glucose metabolism. In accordance with this, administration of nonmetabolizable 3OMG induces the first phase of elevation of MARCKS, while the second elevation of MARCKS is not observed. Once again, glucose evokes a rapid phosphorylation of MARCKS by a

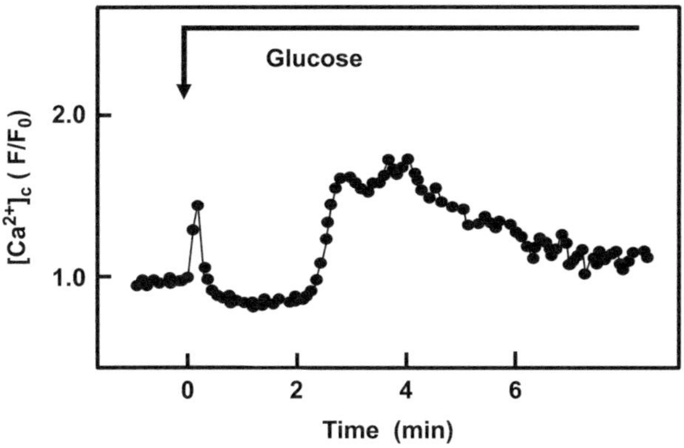

Fig. 4.3 Effect of Glucose on $[Ca^{2+}]_s$ [21]. PM-Cameleon-expressing MIN6 cell was stimulated by 16.7 mM glucose and changes in $[Ca^{2+}]_s$ monitored

mechanism independent of glucose metabolism. It should be noted that the rapid activation of PKC occurs when $[Ca^{2+}]_c$ is reduced by glucose. This seems somewhat peculiar since conventional PKC is activated by Ca^{2+} and diacylglycerol (DAG) [19]. We wondered whether we failed to detect initial changes in $[Ca^{2+}]_c$ induced by glucose. Several possibilities are considered. First, glucose may induce a tiny change in $[Ca^{2+}]_c$, which cannot be detected by the conventional method. Second, glucose increases $[Ca^{2+}]$ in some compartment of the cell, for example, in the subplasma membrane area. Third, fluxes of Ca^{2+} are balanced so that $[Ca^{2+}]$ in the balk of cytosol is not changed significantly.

Considering these possibilities, we decided to detect tiny changes in $[Ca^{2+}]_c$ in β-cells and developed a new method to monitor subplasma membrane $[Ca^{2+}]$ concentration ($[Ca^{2+}]_s$) using ultra-sensitive Ca^{2+} indicator yellow cameleon-nano15 (YC-nano15) [20]. We targeted yellow cameleon-nano15 to the plasma membrane in order to monitor $[Ca^{2+}]_s$ in a very sensitive way [21]. Using this method, we measured changes in $[Ca^{2+}]_s$ in a glucose-stimulated β-cell. As depicted in Fig 4.3, addition of glucose evokes triphasic changes in $[Ca^{2+}]_c$ [21]. Thus, glucose induces an immediate sharp peak of $[Ca^{2+}]_s$. This sharp peak of $[Ca^{2+}]_s$ is followed by a sustained reduction of $[Ca^{2+}]_s$, which lasts a min or more. Then, there is a large oscillatory elevation of $[Ca^{2+}]_c$, which lasts as long as the glucose concentration is high. The rapid peak of $[Ca^{2+}]_s$ has never been detected before. In fact, if Fura-2 is loaded in YC-nano15-expressing cell, the rapid peak is not observed [21]. Presumably, Ca^{2+}-chelating activity of Fura-2 may have abolished tiny changes in $[Ca^{2+}]$ in Fura-2-loaded cells. In any case, glucose evokes an immediate elevation of $[Ca^{2+}]_s$ in β-cells. Among three phases of changes in $[Ca^{2+}]_s$, rapid transient and sustained reduction are independent of glucose metabolism. Thus, addition of mannoheptulose abolished third oscillatory elevation of $[Ca^{2+}]_s$, whereas initial transient and subsequent sustained decrease in $[Ca^{2+}]_s$ were observed in the presence of mannoheptulose. In addition, addition of nonmetabolizable glucose analog 3OMG induced a rapid transient and subsequent sustained decrease in $[Ca^{2+}]_s$. Unlike glucose,

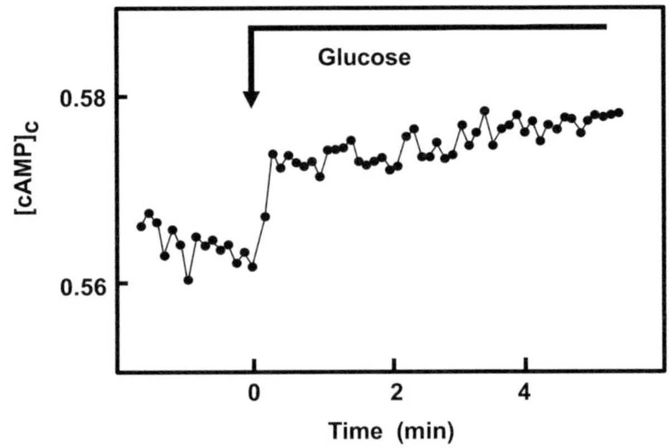

Fig. 4.4 Effect of Glucose on $[cAMP]_c$ [21]. EPac-1-camps-expressing MIN6 cell was stimulated by 16.7 mM glucose and changes in $[cAMP]_c$ were monitored

3OMG did not induce oscillatory elevation of $[Ca^{2+}]_s$. Consequently, the first rapid peak and subsequent reduction of $[Ca^{2+}]_s$ are independent of glucose metabolism, whereas oscillatory elevation of $[Ca^{2+}]_s$ is dependent on glucose metabolism. It is noteworthy that the initial sharp peak and sustained reduction of $[Ca^{2+}]_s$ are blocked by an inhibitor of Gq and an inhibitor of phospholipase C (PLC) [21], suggesting that these two kinds of signals are generated by receptor-mediated activation of Gq and PLC. In contrast, the third oscillatory elevation of $[Ca^{2+}]_s$ may be dependent on K_{ATP} channel and voltage-dependent calcium channel.

It is well known that a high concentration of glucose elevates cyclic AMP (cAMP) in pancreatic β-cells [22, 23]. Elevation of intracellular cAMP concentration ($[cAMP]_c$) has been thought to be due to elevation of $[Ca^{2+}]_c$, which is dependent on glucose metabolism, since calcium-dependent adenylyl cyclase is expressed in these cells. If this is the case, it would be expected that elevation of $[cAMP]_c$ would be observed after a certain lag period. To examine this, we developed a sensitive method to monitor the subplasma membrane concentration of cAMP, ($[cAMP]_s$) using a cAMP indicator Epac1-camp bound to the plasma membrane [21].

Using this method, we monitored changes in $[cAMP]_s$ in glucose-stimulated β-cell. As shown in Fig 4.4, administration of a high concentration of glucose induced a rapid monophasic elevation of $[cAMP]_s$. Elevation of $[cAMP]_c$ was observed immediately after the addition of glucose. This response of $[cAMP]_s$ is quite rapid and, more importantly, is not affected by the addition of mannoheptulose, an inhibitor of glucokinase [21]. Consequently, the glucose-mediated rapid elevation of $[cAMP]_s$ is independent of glucose metabolism. In accordance with this notion, 3OMG, a nonmetabolizable glucose analog, evoked a rapid elevation of $[cAMP]_s$ in β-cells [21]. This rapid response of $[cAMP]_s$ may have been due to the activation of a cell-surface receptor since glucose-mediated increase in $[cAMP]_s$ is inhibited by transducting a gene of dominant-negative mutant of G_s, a GTP-binding protein activating adenylyl cyclase [21]. It seem quite likely that glucose evoked a rapid $[cAMP]_s$ response by receptor-mediated activation of G_s.

The above results indicate that in addition to the well-known metabolism-dependent Ca^{2+} signal, that is, K_{ATP} channel-dependent Ca^{2+} entry, glucose produces rapid signals, and it is not dependent on glucose metabolism. Rather, glucose may activate a cell-surface receptor and generate immediate Ca^{2+} and cAMP signals by activating G proteins, Gq and Gs. We named this putative cell-surface receptor glucose-sensing receptor (GSR) [16]. Since GSR is activated by glucose and non-metabolizable 3OMG, it resembles the sweet taste receptor in the tongue. The sweet taste receptor is expressed in the taste cells of the tongue and is thought to be a heterodimer of two subunits of the type 2 and type 3 taste receptor 1 family (T1R2 and T1R3, respectively) [24, 25]. Both T1R2 and T1R3 are also expressed in pancreatic β-cells [17]. Regarding the expression levels of T1R2 and T1R3, T1R3 is abundantly expressed in β-cells while the expression of T1R2 is low and, in fact, undetectable at the protein level [26]. Hence, T1R3 is a dominant subunit expressed in β-cells. A key question is whether or not T1R3 is involved in the action of glucose. To address this question, we first used lactisole, an inhibitor of T1R3 [27]. In the presence of lactisole, glucose-induced elevation of $[Ca^{2+}]_s$ was altered [21, 28]. Thus, glucose did not induce the rapid peak of $[Ca^{2+}]_s$. Likewise, glucose-induced reduction of $[Ca^{2+}]_s$ was not observed. In addition, the large oscillatory elevation of $[Ca^{2+}]_s$ was reduced in the presence of lactisole. Similar results are obtained when β-cells obtained from T1R3-knock out mice are stimulated by glucose [21]. Thus, in T1R3-deleted β-cells, glucose does not cause an immediate elevation of $[Ca^{2+}]_s$ nor sustained reduction of $[Ca^{2+}]_s$. Additionally, oscillatory elevation of $[Ca^{2+}]_s$ is delayed and the magnitude of the response is reduced in β-cells obtained from T1R3-deleted mice. These results indicate that T1R3 is necessary for the rapid action of glucose on $[Ca^{2+}]_s$. In addition, reduction of $[Ca^{2+}]_s$ caused by glucose requires the function of T1R3. These two signals are mediated by GSR, and T1R3 may be a necessary component of the receptor. Moreover, delayed oscillatory elevation of $[Ca^{2+}]_s$ also partly requires the function of T1R3. Considering the fact that the oscillatory elevation is dependent on glucose metabolism and is due to entry of calcium via the voltage-dependent Ca^{2+} channel, it is reasonable to speculate that GSR may modulate glucose metabolism and/or directly channel functions. In this regard, we have shown that glucose promotes its own metabolism and facilitates generation of ATP by activating GSR [15]. Activation of GSR thus further increases ATP production, further inhibits the K_{ATP} channel, causes further depolarization, and promotes Ca^{2+} entry. Collectively, glucose does not only evoke metabolism-dependent Ca^{2+} signal, but it also generates receptor-mediated Ca^{2+} and cAMP signals [21]. In particular, glucose induces rapid signals by activating a cell-surface receptor GSR. The role of GSR in the action of glucose should be further investigated.

References

1. Rasmussen H, Zawalich KC, Ganesan S, Calle R, Zawalich WS. Physiology and pathophysiology of insulin secretion. Diabetes Care. 1990;13:655–66.
2. Komatsu M, Takei M, Ishii H, Sato Y. Glucose-stimulated insulin secretion: a newer perspective. J Diabetes Invest. 2013;4:511–6.
3. Dean PM, Matthews EK. Electrical activity in pancreatic islet cells. Nature. 1968;219:389–90.
4. Cook DL, Hales CN. Intracellular ATP directly blocks K+ channels in pancreatic B-cells. Nature. 1984;311:271–3.
5. Ashcroft FM, Harrison DE, Ashcroft SJ. Glucose induces closure of single potassium channels in isolated rat pancreatic beta-cells. Nature. 1984;312(5993):446–8.
6. Rutter GA, Pullen TJ, Hodson DJ, Martinez-Sanchez A. Pancreatic β-cell identity, glucose sensing and the control of insulin secretion. Biochem J. 2015;466:203–18.
7. Matschinsky FM. Glucokinase as glucose sensor and metabolic signal generator in pancreatic β-cells and hepatocytes. Diabetes. 1990;39:647–52.
8. Henquin JC. Regulation of insulin secretion: a matter of phase control. and amplitude modulation. Diabetologia. 2009;52:738–51.
9. Grapengiesser E, Gylfe E, Hellman B. Glucose induced oscillation of cytoplasmic Ca^{2+} in the pancreatic β-cell. Biochem Biophys Res Commun. 1988;151:1299–394.
10. Kennedy ED, Rizzuto R, Theler JM, Pralong WF, Bastianutto C, Pozzan T, Wollheim CB. Glucose-stimulated insulin secretion correlates with changes in mitochondrial and cytosolic Ca^{2+} in aequorin-expressing INS-1 cells. J Clin Invest. 1996;98:2524–38.
11. Thore S, Dyachok O, Gylfe E, Tengholm A. Feedback activation of phospholipase C via intracellular mobilization and store-operated influx of Ca^{2+} in insulin-secreting beta-cells. J Cell Sci. 2005;118:4463–71.
12. Roe MW, Merts RJ, Lancaster ME, Worley JF, Dukes ID. Thapsigargin inhibits the glucose-induced decrease of intracellular Ca^{2+} in mouse islets of Langerhans. Am J Phys. 1994;266:E852–62.
13. Tengholm A, Hellman B, Gylfe E. The endoplasmic reticulum is a glucose-modulated high-affinity sink for Ca^{2+} in mouse pancreatic β-cells. J Physiol. 2001;530:533–40.
14. Suzuki Y, Zhang H, Saito N, Kojima I, Urano T, Mogami H. GLP-1 activates protein kinase C through Ca^{2+}-dependent activation of phospholipase C in insulin-secreting cells. J Biol Chem. 2006;281:28499–507.
15. Nakagawa Y, Ohtsu Y, Nagasawa M, Shibata H, Kojima I. Glucose promotes its own metabolism by acting on the cell-surface glucose-sensing receptorT1R3. Endocr J. 2014;61:119–31.
16. Kojima I, Nakagawa Y, Ohtsu Y, Hamano K, Medina J, Nagasawa M. Return of the glucoreceptor: glucose activates the glucose-sensing receptor T1R3 and facilitates metabolism in pancreatic β-cells. J Diabetes Invest. 2015;6:256–63.
17. Nakagawa Y, Nagasawa M, Yamada S, Hara A, Mogami H, Nikolaev VO, Lohse MJ, Shigemura N, Ninomiya Y, Kojima I. Sweet taste receptor expressed in pancreatic β-cells activates the calcium and cyclic AMP signaling systems and stimulates insulin secretion. PLoS One. 2009;4:e5106.
18. Suzuki Y, Zhang H, Saito N, Kojima I, Urano T, Mogami H. Glucagon-like peptide 1 activates protein kinase C through Ca^{2+}-dependent activation of phospholipase C in insulin-secreting cells. J Biol Chem. 2006;281:28499–507.
19. Sjöholm A, Zhang Q, Welsh N, Hansson A, Larsson O, Tally M, Berggren PO. Rapid Ca^{2+} influx and diacylglycerol synthesis in growth hormone-mediated islet beta-cell mitogenesis. J Biol Chem. 2000;14:21033–40.
20. Horikawa K, Yamada Y, Matsuda T, Kobayashi K, Hashimoto M, Matsu-ura T, Miyawaki A, Michikawa T, Mikoshiba K, Nagai T. Spontaneous network activity visualized by ultrasensitive Ca^{2+} indicators, yellow Cameleon-Nano. Nat Methods. 2010;7:729–32.

21. Nakagawa Y, Nagasawa M, Medina J, Kojima I. Glucose evokes rapid Ca^{2+} and cyclic AMP signals by activating the cell-surface glucosesensing receptor in pancreatic β-cells. PLoS One. 2015;10:e0144053.
22. Charles M, Franska R, Schmid F, Forshan P, Grodsky G. Adenosine 3′, 5′-monophosphate in pancreatic islets: glucose-induced insulin release. Science. 1973;179:569–71.
23. Furman B, Ong WK, Pyne NJ. Cyclic AMP signaling in pancreatic islets. Adv Exp Med Biol. 2010;654:281–304.
24. Nelson G, Hoon MA, Chandrasekar J, Zhang Y, Ryba NJP, Zuker C. Mammalian sweet taste receptors. Cell. 2001;106:381–90.
25. Temussi P. The sweet taste receptor: a single receptor with multiple sites and modes of interaction. Adv Food Nutr Res. 2007;53:199–239.
26. Medina A, Nakagawa Y, Ma J, Hamano K, Akimoto T, Ninomiya Y. Kojima. Expression of the glucose-sensing receptor T1R3 in pancreatic islet: changes in the expression levels in various nutritional and metabolic states. Endocr J. 2014;61:797–895.
27. Johnson C, Birch GG, MacDougall DB. The effect of the sweetness inhibitor 2 (-4-methoxyphenoxy) propanoic acid (sodium salt) (Na-PMP) on the taste of bitter–sweet stimuli. Chem Senses. 1994;19:349–58.
28. Hamano K, Nakagawa Y, Ohtsu Y, et al. Lactisole inhibits the glucosesensing receptor in mouse pancreatic β-cells. J Endocrinol. 2015;226:57–66.

Chapter 5
The Role of the Glucose-Sensing Receptor in Glucose-Induced Insulin Secretion in Pancreatic β-Cells

Yuko Nakagawa and Johan Medina

Abstract Glucose activates the glucose-sensing receptor and induces rapid intracellular signals in pancreatic β-cells. When the glucose-sensing receptor is blocked by an inhibitor of T1R3 or deletion of the T1R3 gene, glucose-induced insulin secretion (GIIS) is significantly reduced. In perifusion system, both first and second phases of GIIS are attenuated by the inhibition of the glucose-sensing receptor. Collectively, the glucose-sensing receptor is involved in both rapid and sustained action of glucose. Indeed, activation of the receptor by either artificial sweeteners or nonmetabolizable glucose analog increases ATP levels in β-cells. Furthermore, inhibition of the glucose-sensing receptor attenuates glucose-induced increase in ATP. These results indicate that activation of the glucose-sensing receptor promotes glucose metabolism and thereby augments ATP production in β-cells. Thus, glucose first acts on the cell-surface glucose-sensing receptor and primes the metabolic pathway of glucose. Glucose then enters β-cells and is metabolized through already activated metabolic pathway. The receptor pathway and the metabolic pathway act coordinately to stimulate insulin secretion.

Keywords Insulin secretion · Pancreatic β-cell · Glucose · Glucose-sensing receptor · Glucose metabolism

5.1 Introduction

Glucose is a principal fuel in the body that acts as a critical stimulator of insulin secretion from pancreatic β-cells. The mechanism by which glucose stimulates insulin secretion is an important issue from basic and clinical points of view, and it has been investigated for several decades by many researchers. It is now thought that glucose exerts its action though its metabolism [1]. Thus, glucose enters β-cells,

Y. Nakagawa (✉) · J. Medina
Institute for Molecular and Cellular Regulation, Gunma University, Maebashi, Japan
e-mail: nakayuko@gunma-u.ac.jp; j.medina@uq.edu.au

I. Kojima (ed.), *Glucose-sensing Receptor in Pancreatic Beta-cells*,
https://doi.org/10.1007/978-981-13-0002-8_5

is catalyzed by glucokinase and downstream glycolytic enzymes and in mitochondria, and the resultant increase in the ATP/ADP ratio causes closure of the ATP-sensitive potassium channel (K_{ATP} channel). Resultant depolarization leads to opening of the voltage-gated calcium channel, and Ca^{2+} enters the cell. These sequences of events are dependent on glucose metabolism, and when glucokinase is blocked by mannoheptulose, for example, both Ca^{2+} response and insulin secretion do not take place [1].

5.2 Sweet Taste Receptor in Pancreatic β-Cells

We found recently that glucose generates rapid intracellular signals in pancreatic β-cells, which are not dependent on glucose metabolism [2]. Thus, glucose evokes rapid Ca^{2+} signals that are not blocked by mannoheptulose, an inhibitor of glucokinase [3]. Also, these rapid Ca^{2+} signals are reproduced by adding 3-*O*-methylglucose (3OMG), a glucose analog not catalyzed in β-cells. Furthermore, these Ca^{2+} signals are attenuated by a Gq inhibitor and an inhibitor of phospholipase C (PLC) [3]. Likewise, glucose evokes a rapid elevation of cyclic AMP (cAMP), which is not affected by mannoheptulose but is reproduced by 3OMG. In addition, glucose-induced elevation of cAMP is attenuated by transduction of a gene encoding dominant-negative mutant of G_s [3]. All these results suggest that glucose activates a cell-surface receptor and produces rapid intracellular signals by activating G proteins. We named this receptor glucose-sensing receptor (GSR) [4].

What is the molecular nature of the GSR? In this regard, Niki and colleagues showed three decades ago that an inhibitor of sweet taste sensation, *p*-nitrophenyl-D-glucopyranoside (PNP-Glu), attenuated glucose-induced insulin secretion [5]. In contrast, PNP-Glu did not affect insulin secretion induced by a high concentration of potassium. Their results raised a possibility that the sweet taste receptor or related molecules may be expressed in pancreatic β-cells and mediates the action of glucose.

The sweet taste receptor is expressed in taste cells of the taste bud in the tongue. In 2001, the molecular nature of the sweet taste receptor was revealed by molecular cloning and functional analyses [6, 7]. It is now generally thought that the sweet taste receptor is a heterodimer comprised of the two members of the taste receptor-1 (T1R) family, type 2 and type 3 T1R (T1R2 and T1R3, respectively). Both T1R2 and T1R3 are members of the class C G protein-coupled receptor (GPCR) superfamily. We speculated that the sweet taste receptor subunits are expressed in pancreatic β-cells and mediate the action of glucose [2]. Indeed, reverse-transcription PCR (RT-PCR) revealed that mRNA for T1R2 and T1R3 is expressed in mouse pancreatic islets [8]. Also, in MIN6 cells, a glucose-responsive mouse β-cell line [9], mRNA for T1R2 and T1R3 has been detected. Furthermore, immunoreactive T1R3 has been detected in the core of mouse islets and in MIN6 cells [8]. Based on these data, we initially thought that the canonical sweet taste receptor, i.e., a heterodimer of T1R2 and T1R3 (T1R2/T1R3), is expressed in pancreatic β-cells [8]. The function of the sweet taste receptor can be assessed by activating it using artificial sweeteners. Indeed, when sucralose, a potent artificial sweetener, is added, insulin secretion is augmented [8]. Likewise, other sweeteners, for example, acesulfame-potassium, saccharin, and glycyrrhizin all increase insulin secretion [8]. Therefore, the "sweet taste receptor" in β-cells seems to be functional. It should be mentioned that relatively high concentrations of artificial sweeteners, for example, 10–20 mM, are needed to stimulate insulin secretion. We assessed the signal transduction pathways activated by the "sweet taste receptor" using MIN6 cells [8]. Because of the complex structure of the taste buds, the signal transduction pathway activated by the sweet taste receptor has not been investigated thoroughly in taste cells of the tongue. β-cells may provide a good system to study the signal transduction of the sweet taste receptor. In fact, when MIN6 cells are stimulated by sucralose, both cytoplasmic Ca^{2+} concentration ($[Ca^{2+}]_c$) and cytoplasmic cAMP concentration ($[cAMP]_c$) are elevated [8]. Sucralose also activates protein kinase C (PKC) [8]. Consequently, both Ca^{2+} and cAMP messenger systems are activated by an artificial sweetener sucralose. This is an interesting feature of the "sweet taste receptor" in β-cells. It should be noted that the effect of sucralose on $[Ca^{2+}]_c$ is attenuated by gurmarin [8], an inhibitor of the sweet taste receptor, and lactisole [10], an inhibitor of T1R3. Knockdown of T1R3 also attenuates the effect of sucralose [11]. Hence, sucralose exerts its effect by acting on the T1R3-containing receptor, presumably the "sweet taste receptor." An intriguing aspect of the "sweet taste receptor" in β-cells is that it produces a variety of patterns of signals. When MIN6 cells are stimulated by other sweeteners, considerably different types of signals are produced compared to those evoked by sucralose [12]. For example, when cells are stimulated by saccharin, a classical artificial sweetener, $[cAMP]_c$, is increased, whereas $[Ca^{2+}]_c$ is not affected. In contrast, glycyrrhizin, a natural sweetener derived from licorice, increases $[Ca^{2+}]_c$ but not $[cAMP]_c$ [12]. Five or more sweeteners examined so far induce different patterns of intracellular signals presumably by activating different transducers and/or effectors [12]. The signal transduction mechanism activated by the "sweet taste receptor" is quite unique: different types of agonists generate distinct patters of signals. In this regard, these ligands may act as biased agonists. In fact, the sweet taste receptor in the tongue is activated by numerous natural or artificial sweet substances with a variety of chemical structures (7). Those ligands have multiple binding sites in the receptor molecule [7]. Ligands binding to distinct binding sites may activate different sets of responses. It is also possible that the "sweet taste receptor" in β-cells is not a single receptor molecule but is comprised of multiple types of receptor molecules. Then, it is possible that multiple ligands produce multiple types of signals by binding to different subtypes of the receptor. In this situation, these receptor molecules should contain T1R3 since inhibition or deletion of T1R3 attenuates the receptor functions. If T1R3 is a component of the receptor, it is unlikely that five or more types of receptors exist. Collectively, at least some of the ligands act as biased agonists.

As mentioned above, GSR responds to physiological concentrations of glucose and produces immediate signals in β-cells [3]. A critical question is whether or not GSR is identical to the canonical sweet taste receptor expressed in the tongue, i.e.,

T1R2/T1R3 heterodimer. In other words, whether or not the canonical sweet taste receptor is activated by physiological concentrations of glucose in the plasma is a critical question. The answer may be No. In the taste buds, it is known that glucose is much less potent than sucrose in activating the sweet taste receptor [7]. It is therefore unlikely that the canonical sweet taste receptor, i.e., T1R2/T1R3, functions as GSR. An important point is that T1R3 is required for the function of GSR [3]. Consequently, it is reasonable to speculate that GSR may be a dimer containing T1R3, which is distinct from T1R2/T1R3. We therefore re-examined the expression of T1Rs in pancreatic β-cells in more detail. When we measured the expression of T1R2 and T1R3 by quantitative RT-PCR, the expression of T1R3 was abundant whereas the expression of T1R2 was far less [13]. In fact, when we measured the protein expression by either immunohistochemistry or immunoblotting, T1R3 was detected abundantly in β-cells while T1R2 was undetectable [13]. These results indicate that in pancreatic β-cells, the expression of T1R2 is negligible and the canonical sweet taste receptor, T1R2/T1R3, is therefore a very minor component, if any.

5.3 Glucose-Sensing Receptor in Pancreatic β-Cells

The above results indicate that the receptor activated by glucose and sweeteners in β-cells may be slightly different from the canonical sweet taste receptor, T1R2/T1R3. Then what is it? Since class C GPCR functions as a dimer [15], and T1R3 is a major T1R in β-cells, it is reasonable to speculate that the putative receptor may be T1R3-containing dimer other than T1R2/T1R3. Theoretically, candidate receptors are a homodimer of T1R3 (T1R3/T1R3) and a heterodimer of T1R3 and X, where X is a class C GPCR expressed in pancreatic β-cells. To this end, we first examined whether or not T1R3/T1R3 functions as GSR. As we showed previously, T1R3/T1R3 is able to function as a signaling receptor [14]. Using HEK cells stably expressing T1R3 (HEK-T1R3 cells), we examined whether or not glucose increases $[Ca^{2+}]_c$. As depicted in Fig. 5.1, basal $[Ca^{2+}]_c$ is stable and a high concentration of glucose induces an immediate elevation of $[Ca^{2+}]_c$. This elevation is only transient and is followed by a small but sustained reduction of $[Ca^{2+}]_c$. Noteworthy is the fact that this pattern of $[Ca^{2+}]_c$ response is quite similar to that observed in β-cells stimulated by nonmetabolizable glucose analog 3OMG [3]. T1R3/T1R3 is therefore capable of producing this pattern of $[Ca^{2+}]_c$ response. This is an intriguing feature of T1R3/T1R3, and such responses are not observed in cells expressing ordinary class A GPCR. Interestingly, glucose-induced changes in $[Ca^{2+}]_c$ are attenuated by a Gq inhibitor and an inhibitor of PLC, suggesting that both elevation and reduction of $[Ca^{2+}]_c$ are coupled and are mediated by Gq and subsequent PLC. This is the reason reduction of $[Ca^{2+}]_c$ is observed after the transient elevation of $[Ca^{2+}]_c$. A possible interpretation is that Ca^{2+} is taken up to ER, and simultaneously, Ca^{2+} entry is inhibited by unknown reasons. When various concentrations of glucose are tested, however, a relatively high concentration of glucose, for example 25 mM, is required to elicit $[Ca^{2+}]_c$ responses in HEK-T1R3 cells. This means that although T1R3/T1R3

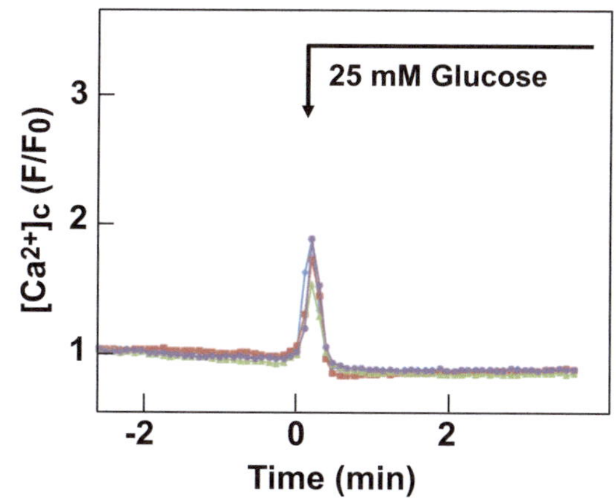

Fig. 5.1 Effect of glucose on $[Ca^{2+}]_c$ in HEK-T1R3 cells. Fluo-8-loaded HEK-T1R3 cells were stimulated by 25 mM glucose and changes in $[Ca^{2+}]_c$ were measured

is an interesting candidate for GSR, glucose sensitivity is not high enough. Another candidate receptor needs to be considered.

As mentioned above, GSR could be a heterodimer of T1R3 and X, where X is a class C GPCR expressed in pancreatic β-cells. The most interesting candidate for X is the calcium-sensing receptor (CaSR), which is expressed abundantly in pancreatic β-cells [15]. CaSR is identified as a cell-surface receptor detecting the changes in extracellular Ca^{2+} concentrations. It is expressed in parathyroid cells, renal tubular cells, and bone cells, three major organs regulating calcium metabolism in the body. Subsequently, the molecular structure of CaSR has been identified by molecular cloning, and the results show that CaSR belongs to the class C GPCR family acting as a homodimer. It is a multifunctional receptor and is in fact activated not only by Ca^{2+} but also by many other compounds including various amino acids and cationic compounds [16]. We established a stable HEK cell line expressing CaSR [17] and examined whether or not glucose is able to activate the receptor. We found that many sweet substances including sugars and artificial sweeteners activated CaSR [17]. Indeed, an addition of glucose induced a rapid increase in $[Ca^{2+}]_c$. The rapid Ca^{2+} transient was followed by sustained reduction of $[Ca^{2+}]_c$ [17]. This pattern of $[Ca^{2+}]_c$ response is rather unique but resembles that induced by 3OMG in β-cells. Glucose-induced biphasic changes in $[Ca^{2+}]_c$ were inhibited by inhibitors of Gq and PLC. Hence, the unique pattern of $[Ca^{2+}]_c$ response is dependent on Gq-mediated activation of PLC. To our surprise, CaSR is very sensitive to glucose and as low as 5 mM of glucose-induced the maximal response. Thus, when ambient glucose concentration was raised from 3 to 5 mM, a marked transient of $[Ca^{2+}]_c$ was observed, which was followed by sustained reduction of $[Ca^{2+}]_c$. CaSR homodimer is too sensitive to function as a physiological receptor for glucose. However, CaSR may be a good candidate receptor that dimerizes with T1R3.

We then examined the effect of glucose in stable HEK cell line expressing both T1R3 and CaSR (HEK-T1R3/CaSR cells). In these cells, formation of a heterodimer of T1R3 and CaSR was detected. As shown in Fig. 5.2, oscillation of $[Ca^{2+}]_c$ was observed in an unstimulated condition. Addition of a high concentration of

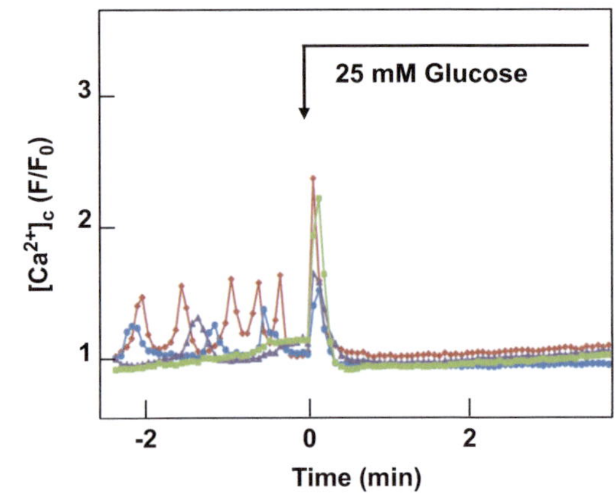

Fig. 5.2 Effect of glucose on $[Ca^{2+}]_c$ in HEK-T1R3/CaSR cells. Fluo-8-loaded HEK-T1R3/CaSR cells were stimulated by 25 mM glucose and changes in $[Ca^{2+}]_c$ were measured

glucose induces biphasic response of $[Ca^{2+}]_c$ in HEK-T1R3/CaSR cells: a rapid transient peak followed by a small but sustained decrease in $[Ca^{2+}]_c$. Again, this pattern of changes in $[Ca^{2+}]_c$ is similar to that observed in β-cells stimulated by 3OMG [3]. In these cells, 7 to 8 mM of glucose is able to induce $[Ca^{2+}]_c$ response. A heterodimer of T1R3/CaSR is therefore an interesting candidate for GSR. In MIN6 cells, NPS-2143, an inhibitor of CaSR, blocks the $[Ca^{2+}]_c$ response induced by 3OMG. Similarly, lactisole, an inhibitor of T1R3 blocks $[Ca^{2+}]_c$, responses induced by 3OMG. Consequently, both T1R3 and CaSR are required for the action of 3OMG on $[Ca^{2+}]_c$. A heterodimer of T1R3 and CaSR is the best candidate for the GSR.

5.4 Role of GSR in Glucose-Induced Insulin Secretion

Glucose activates cell-surface GSR and evokes rapid signals including elevation of $[Ca^{2+}]_c$, activation of protein kinase C (PKC), and elevation of $[cAMP]_c$. Elevation of Ca^{2+} concentration, especially in the subplasma membrane area, is a trigger for exocytosis of insulin granules, and both cAMP and diacylglycerol act as allosteric modulators of exocytosis and increase the Ca^{2+}-sensitivity of exocytosis. It is thus quite likely that intracellular signals evoked by activation of GSR facilitate insulin secretion. In fact, inhibition of T1R3 by adding lactisole significantly reduces glucose-induced insulin secretion [10]. Since GSR signals are rapid, we initially speculated that the GSR signals are important for the first phase of glucose-induced insulin secretion. We tested this idea by perifusion experiments. When we compared glucose-induced insulin secretion in a perifusion system using islets obtained from normal and T1R3-knockout mice, the results were slightly different from what we expected. As shown in Fig. 5.3, both first and second phases of glucose-induced insulin secretion were significantly reduced in islets obtained from T1R3-knockout mice. These results are rather unexpected but coincide with the results obtained by measuring $[Ca^{2+}]_c$ response [3]. Inhibition of T1R3 reduces not only the rapid

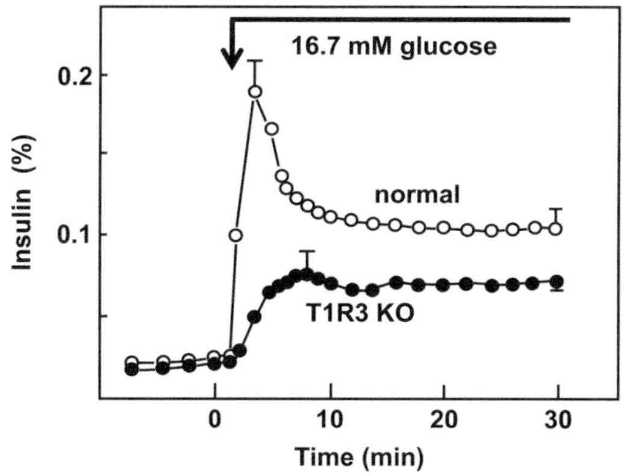

Fig. 5.3 Effect of glucose on insulin secretion in islets from normal and T1R3 knockout mice. Islets obtained from normal (○) and T1R3-knockout (●) mice were stimulated by 16.7 mM glucose, and insulin secretion was measured in a perifusion system

response but also subsequent oscillatory $[Ca^{2+}]_c$ responses observed several minutes after the stimulation by a high concentration of glucose in β-cells [3]. The oscillatory elevation of $[Ca^{2+}]_c$ has been thought to be due to Ca^{2+} entry caused by the inhibition of K_{ATP} channel. These results indicate that GSR not only induces rapid signals, but it also modulates sustained action of glucose in β-cells. How does GSR modulate the long-term action of glucose? Since glucose metabolism is important for sustained action of glucose, we speculated that the GSR signals modulate glucose metabolism in β-cells. To test this idea, we monitored changes in intracellular ATP ($[ATP]_c$) using luciferase-expressing MIN6 cells [11]. We first examined changes in $[ATP]_c$ induced by various concentration of glucose. As depicted in Fig. 5.4a, glucose induced biphasic elevation of $[ATP]_c$ in MIN6 cells, and the effect of glucose was dose-dependent. Conversely, $[ATP]_c$ was rapidly reduced by inhibition of mitochondrial function by either dinitrophenol or 2-cycrohexen-1-one, indicating that changes in $[ATP]_c$ in MIN6 cells could be monitored. Using this system, we first tested whether or not activation of GSR by sucralose affected $[ATP]_c$. To our surprise, addition of sucralose markedly increased $[ATP]_c$ in MIN6 cells (Fig. 5.4b). This experiment was performed in the presence of 5.5 mM of glucose in the incubation medium. It should be noted that sucralose is an artificial sweetener and therefore does not enter β-cells nor serve as a fuel. Nevertheless, sucralose markedly elevated $[ATP]_c$ in MIN6 cells. The effect of sucralose was also confirmed by monitoring the ATP/ADP ratio by using Percival [11]. This effect of sucralose is not simply due to the elevation of $[Ca^{2+}]_c$ or $[cAMP]_c$ induced by sucralose since addition of muscarinic agonist carbachol or glucagon-like peptide-1, which increases $[Ca^{2+}]_c$ or $[cAMP]_c$, respectively, did not affect $[ATP]_c$ [11]. Furthermore, the effect of sucralose on $[ATP]_c$ was observed even in the absence of ambient glucose. When MIN6 cells were incubated for 60 min in medium containing no glucose, the basal level of $[ATP]_c$ was reduced. However, sucralose was still able to increase $[ATP]_c$ in this condition. Since glycogen content, if any, should be severely reduced in this condition, the result suggests that sucralose is able to increase ATP by not utilizing glucose-6-phosphate as a substrate. This effect of sucralose is due to the activation of GSR since nonmetabolizable analog of glucose 3OMG reproduced the sucralose

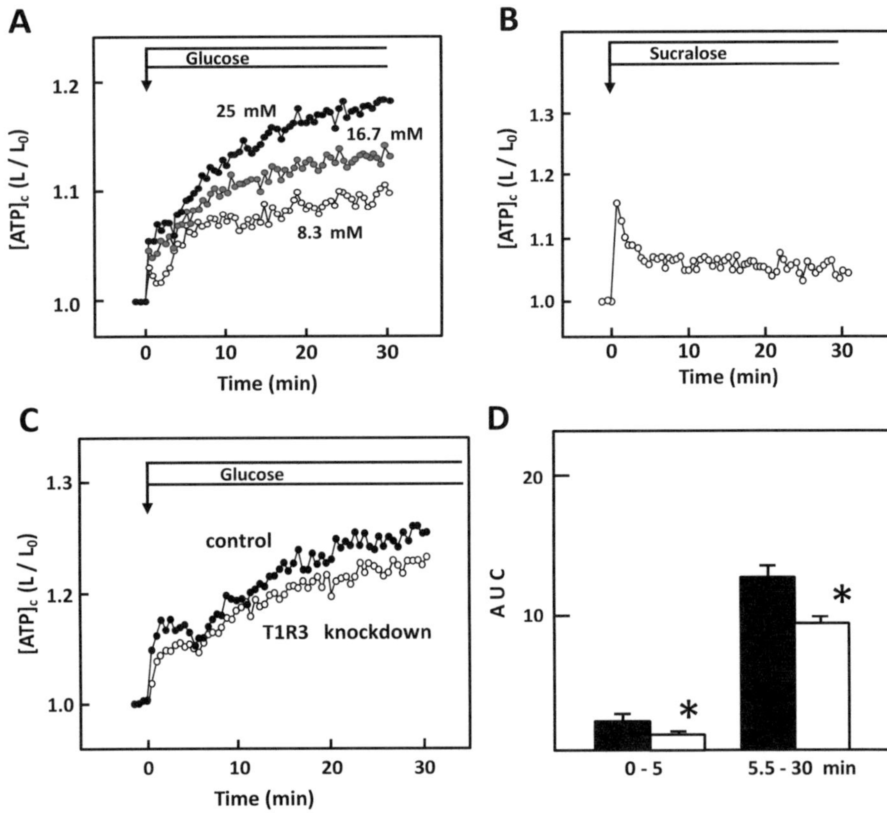

Fig. 5.4 Effect of glucose and sucralose on $[ATP]_c$ in luciferase- expressing MIN6 cells [11]. (**a**) Effects of various concentration of glucose on $[ATP]_c$. Luciferase- expressing MIN6 cells were incubated in Krebs-Ringer bicarbonate buffer containing 5.5 mM glucose. Then glucose concentration was raised to 8.3, 16.7, and 25 mM glucose, and changes in $[ATP]_c$ were monitored. (**b**) Effect of sucralose on $[ATP]_c$. Luciferase-expressing MIN6 cells were stimulated by 10 mM sucralose and changes in $[ATP]_c$ were monitored. (**c**) Effect of knock-down of T1R3 on glucose-induced increase in $[ATP]_c$. Luciferase-expressing MIN6 cells and T1R3-knocked down MIN6 cells were stimulated by 25 mM glucose and changes in $[ATP]_c$ were monitored (**d**) Effect of knockdown of T1R3 on glucose-induced in $[ATP]_c$. Experiments were done as in **c** and area under the curve from 0 to 5 and 5.5 to 30 min was calculated

effect. Also, lactisole, an inhibitor of T1R3, attenuated the effect of sucralose [10]. GSR signals thus activate the metabolic pathway and increase $[ATP]_c$. One of the sites of action of the GSR signals is promotion of the metabolism in mitochondria since sucralose and mitochondrial fuel methylsuccinate act synergistically to increase $[ATP]_c$. As mentioned above, 3OMG increases $[ATP]_c$. This means that glucose not only serves as a substrate but it also activates GSR as a ligand and promotes its own metabolism. In fact, the effects of glucose on $[ATP]_c$ and insulin secretion were attenuated by knocking down T1R3 (Fig 5.4c, d) [11]. Collectively, glucose exerts its action by two different mechanisms. Glucose first acts on the

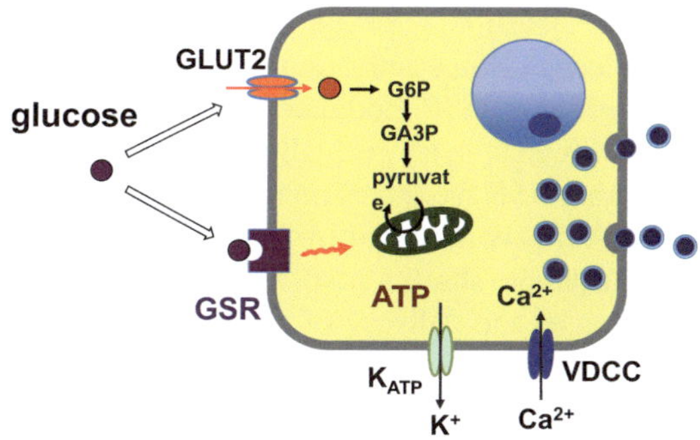

Fig. 5.5 Action of glucose in pancreatic β-cells [11]. Glucose first activates the cell-surface glucose-sensing receptor (GSR) and primes the metabolic pathway. Glucose then enters β-cell and is metabolized through already activated metabolic pathway. The receptor pathway and the metabolic pathway act coordinately to stimulate insulin secretion

cell-surface GSR and activates the receptor. Activation of this receptor causes priming of the metabolic pathway and thereby facilitates glucose metabolism. Then glucose enters β-cells and is catalyzed through the already activated metabolic pathway (Fig. 5.5). Thus, glucose acts as a ligand and also a substrate in β-cells, and it activates two distinct pathways: the GSR pathway and the metabolic pathway. These two pathways merge inside the β-cells, leading to production of more ATP and subsequent inhibition of K_{ATP} channel. These two pathways act coordinately to stimulate insulin secretion [4]. This is a new model showing the mechanism of action of glucose in β-cells. This model is an extension of the "glucoreceptor hypothesis" [2, 4], but it is also includes the "metabolic hypothesis." In fact, in this model, there exists an interaction of the "receptor hypothesis" and the "metabolic hypothesis," and the receptor pathway positively regulates the metabolic pathway. These two pathways act coordinately to exert the action of glucose.

5.5 Future Direction of the GRS Research

Many questions still remain unanswered. First, which step(s) of the glucose metabolism is modulated by the GSR signals? In this regard, glucokinase is a rate-limiting enzyme in the glycolytic pathway, and it has been thought to act as a glucose sensor [18]. Whether or not glucokinase activity is modulated by GSR is an interesting question. There are many steps in the downstream of glucokinase, which could be a target(s) of the GSR. It is necessary to identify the step(s) modulated by the receptor signal. Second, how do the GSR signals stimulate the glucose metabolism? In other words, by what mechanisms does GSR modulate the glucose metabolism? Third, what is the physiological and pathophysiological significance of the GSR pathway?

With regard to the first question, GSR appears to modulate multiple steps in the metabolic pathway of glucose. Since activation of the GSR increases $[ATP]_c$ even in the absence of ambient glucose, GSR modulates the step(s) downstream of glucose-6-phosphate. Indeed, GSR acts synergistically with methylsuccinate to increase $[ATP]_c$ [11], indicating that GSR facilitates metabolism in mitochondria. In fact, the effect of sucralose on $[ATP]_c$ is inhibited by inhibition or deletion of the enzyme involved in the malate-aspartate shuttle. Also, metabolome analyses show that activation of the GSR increases the delivery of substrates to the glycolytic pathway from alanine and glycerol (unpublished observation). Collectively, GSR augments glucose metabolism by acting on the multiple steps in the metabolic pathways. Further studies are clearly needed to determine the precise steps and the regulatory mechanism responsible for promotion of the metabolism. Regarding the second question, we addressed the signal transduction pathways responsible for the elevation of $[ATP]_c$ [19]. Among the signaling pathways activated by the GSR, depolarization of the plasma membrane and resultant entry of Ca^{2+} are important for promotion of glucose metabolism [19]. This is not surprising because there are three dehydrogenases in the mitochondria regulating the TCA cycle. Also, the malate-aspartate shuttle is modulated by Ca^{2+}. In addition, depolarization of the plasma membrane per se may facilitate the metabolism by a yet unknown mechanism [19]. It should be mentioned that the role of the GSR-mediated rapid signals is still unclear. Again, further studies are necessary to identify these mechanisms. Regarding the third question, the GSR signals are necessary for the full action of glucose since inhibition of the GSR function and deletion of the receptor subunit significantly reduce the insulin secretory response to glucose [2]. As mentioned above, both first and second phases of glucose-induced insulin secretion are inhibited without the GSR signal. Since GSR is required for the full action of glucose, it is possible that derangement of the GSR pathway is involved in the pathophysiology of type 2 diabetes. In fact, our results obtained in animal models of diabetes show that the expression of T1R3 is markedly reduced in diabetic animals [13]. This reduction of the expression is due at least partly to prolonged hyperglycemia. When hyperglycemia is corrected by treatment with insulin, the expression of T1R3 is recovered [13]. It is reasonable to speculate that impairment of insulin secretion observed in diabetes is partly due to down-regulation of the GSR by prolonged hyperglycemia. Reduction of the expression of GSR is one of the features of glucotoxicity in diabetes. Elucidation of the regulation of the expression level of the GSR would help with better understanding of the pathophysiology of type 2 diabetes.

References

1. Henquin JC. Regulation of insulin secretion: a matter of phase control and amplitude modulation. Diabetologia. 2009;52:738–51.
2. Kojima I, Medina J, Nakagawa Y. Role of the glucose-sensing receptor in insulin secretion. Diabetes Obes Metab. 2018;19(Suppl 1):1954–62.

3. Nakagawa Y, Nagasawa M, Medina J, Kojima I. Glucose evokes rapid Ca^{2+} and cyclic AMP signals by activating the cell-surface glucose sensing receptor in pancreatic β-cells. PLoS One. 2015;10:e0144053.
4. Kojima I, Nakagawa Y, Ohtsu Y, Hamano K, Medina J, Nagasawa M. Return of the glucoreceptor: glucose activates the glucose-sensing receptor T1R3 and facifitates metabolism in pancreatic β-cells. J Diabetes Investig. 2015;6:256–63.
5. Niki A, Niki H, Hashioka T. Receptors of paraneurons, with special reference to glucoreceptors. Arch Histol Cytol. 1989;52:33–8.
6. Nelson G, Hoon MA, Chandrasekar J, Zhang Y, Ryba NJP, Zuker C. Mammalian sweet taste receptors. Cell. 2001;106:381–90.
7. Temussi P. The sweet taste receptor: a single receptor with multiple sites and modes of interaction. Adv Food Nutr Res. 2007;53:199–239.
8. Nakagawa Y, Nagasawa M, Yamada S, et al. Sweet taste receptor expressed in pancreatic β-cells activates the calcium and cyclic AMP signaling systems and stimulates insulin secretion. PLoS One. 2009;4:e5106.
9. Miyazaki J, Araki K, Yamato E, Ikegami H, Asano T, Shibasaki Y, Oka Y, Yamamura K. Establishment of a pancreatic β cell line that retains glucose-inducible insulin secretion. Endocrinology. 1990;127:126–32.
10. Hamano K, Nakagawa Y, Ohtsu Y, Li L, Medina J, Tanaka Y, Masuda K, Komatsu M, Kojima I. Lactisole inhibits the glucosesensingreceptor in mouse pancreatic β-cells. J Endocrinol. 2015;226:57–66.
11. Nakagawa Y, Ohtsu Y, Nagasawa M, Shibata H, Kojima I. Glucose promotes its own metabolism by acting on the cell-surface glucosesensing receptor T1R3. Endocr J. 2014;61:119–31.
12. Nakagawa Y, Nagasawa M, Mogami H, Lohse M, Ninomiya Y, Kojima I. Multimodal function of the sweet taste receptor expressed in pancreatic β-cells: generation of diverse patterns of intracellular signals by sweet agonists. Endocr J. 2013;60:1191–206.
13. Medina A, Nakagawa Y, Ma J, et al. Expression of the glucose-sensing receptor T1R3 in pancreatic islet: changes in the expression levels in various nutritional and metabolic states. Endocr J. 2014;61:797–895.
14. Masubuchi Y, Nakagawa Y, Ma JH, Kojima I, Shibata H. Novel regulatory function of sweet taste-sensing receptor in adipogenic differentiation of 3T3-L1 cells. PLoS One. 2013;8:e54500.
15. Rasschaert J, Malaisse WJ. Expression of the calcium-sensing receptor in pancreatic islet B-cells. Biochem Biophys Res Commum. 1999;264:615–8.
16. Jensen AA, Bräuner-Osborne H. Allosteric modulation of the calcium-sensing receptor. Curr Neuropharmacol. 2007;5:180–6.
17. Medina J, Nakagawa Y, Nagasawa M, et al. Positive allosteric modulation of the calcium-sensing receptor by physiological concentrations of glucose. J Biol Chem. 2016;291(44):23126–35.
18. Matschinski FM. Glucokinase as glucose sensor and metabolic signal generator in pancreatic b-cells and hepatocytes. Diabetes. 1990;39:647–52.
19. Li LF, Ohtsu Y, Nakagawa Y, Masuda K, Kojima I. Sucralose, an activator of the glucose-sensing receptor, increases ATP by calcium-dependent and -independent mechanisms. Endocr J. 2016;63:715–25.

MIX
Papier aus verantwortungsvollen Quellen
Paper from responsible sources
FSC® C105338

If you have any concerns about our products,
you can contact us on
ProductSafety@springernature.com

In case Publisher is established outside the EU,
the EU authorized representative is:
Springer Nature Customer Service Center GmbH
Europaplatz 3, 69115 Heidelberg, Germany

Printed by Libri Plureos GmbH
in Hamburg, Germany